Gordon H. Orians   Rodolfo Dirzo   J. Hall Cushman (Eds.)

Biodiversity and Ecosystem Processes in Tropical Forests

With 18 Figures

Springer

Prof. dr. Gordon Orians
Dept. of Zoology
University of Washington
Seattle, WA 98195
USA

Prof. Dr. Rodolfo Dirzo
Centro de Ecología
Univesidad Nacional Autónoma de México
AP 70-275
México, 04510 D.F.
México

Prof. Dr. J. Hall Cushman
Dept. of Biology
Sonoma State University
Rohnert Park, CA 94928
USA

ISBN 3-540-59275-X Springer-Verlag Berlin Heidelberg New York

Library of Congress Cataloging – in– Publication Data
Biodiversity and ecosystem processes in tropical forests / Gordon H. Orians, Rodolfo Dirzo, J. Hall Cushmann, editors. p. cm — (Ecological studies ; v. 122) Includes bibliographical references and index.
ISBN 3-540-59275-X
1. Rain forest ecology —Tropics. 2. Biological diversity- Tropics. I. Orians, Gordon H. II. Dirzo, Rodolfo. III.Cushman J. Hall, 1959- . IV. Series.
QH541.5.R27B56 1996   574.5`2642`0913 —dc20

Cover Design: D & P Heidelberg
Typesetting: Camera ready by Dr. Kurt Darms, Bevern
SPIN 10493645   31/3137 5 4 3 2 1 0 - Printed on acid free paper -

Preface

Today, rates of extinction of species are high over much of the earth, but extinction rates of tropical species are probably higher than elsewhere, owing to rapid conversion of tropical forests to other uses, combined with the great species richness of tropical forests. Estimates of deforestation rates vary greatly at both regional and global scales, and statistics often are manipulated because important political and economic consequences are at stake.

Better knowledge of the functioning of tropical forests is needed if biologists are to determine which species and which relationships are vital for the maintenance of the functional properties of tropical forests and the species living in them. Better knowledge can assist targeting limited human and financial resources toward preservation of key species and processes at the locations and scales at which they will be most effective. This book is a contribution to current efforts to provide such knowledge. It is the product of a workshop held at Oaxtepec, Morelos, México, December 4-7, 1993, during which a group of ecologists focused on the roles of species and groups of species in maintaining ecological processes in tropical forests and buffering them in the face of various perturbations. The major product of the workshop is a set of recommendations for future research. These recommendations are based on identifying topics about which critical knowledge is especially lacking. Participants identified species likely to have important influences on ecosystem processes and processes likely to be controlled by relatively few species even in speciose tropical forests. The workshop was supported by a grant from the US National Science Foundation, supplemented by logistical support provided by the Universidad Nacional Autónoma de México. We hope that the results of our deliberations will stimulate research so that future workshops on the same topic will demonstrate significant progress in understanding the role of species richness in the dynamic properties of tropical forests.

Spring 1996

Seattle, Washington, USA   G. H. Orians

Mexico City   R. Dirzo

Rohnert Park, California, USA   J. H. Cushman

CONTENTS

8   Species Richness and Resistance to Invasion....................153
      Marcel Rejmanek

9   The Role of Biodiversity in Tropical Managed
      Ecosystems...173
      Alison Power and Alexander S. Flecker

1 Introduction

Gordon H. Orians[1], Rodolfo Dirzo[2], and J. Hall Cushman[3]

Living organisms on Earth are being subjected to massive disruptions in the form of wholesale exchanges of species among regions, introduction of alien predators and pathogens, overharvesting, habitat destruction, pollution, and, in the future, perhaps also climate change. Changes in land use in the tropics are creating extensive areas of agricultural lands and early successional patches at the expense of late successional and climax communities. Along with these changes are major reductions in population sizes and extinctions of species that depend upon the habitats that are being destroyed. Rates of forest destruction are higher in tropical regions today than elsewhere on Earth (Green and Sussman 1990, Whitmore and Sayer 1992, Wilson 1992, FAO 1993, Skole and Tucker 1993). The consequences of these changes are poorly known. Unfortunately, losses of species that may accompany forest destruction are irretrievable, and habitat loss, although reversible, can be recovered only slowly and with difficulty.

Biologists have directed most of their attention to date towards estimating the extent of species loss, identifying the major causes of those losses, and developing ways to reduce the rate of species extinctions. These efforts have identified habitat destruction as the most important contemporary cause of the estimated extinctions of species. Much less attention has been paid to the consequences of the loss of species in terms of functional proc-

[1] Gordon H. Orians, Department of Zoology, University of Washington, Box 351800, Seattle, Washington 98195, USA

[2] Rodolfo Dirzo, Centro de Ecología, Universidad Nacional Autónoma de México, Apartado Postal 70-275, México 04510 D. F.

[3] J. Hall Cushman, Department of Biology, Sonoma State University, Rohnert Park, California 94928, USA

Ecological Studies, Vol. 122
Orians, Dirzo and Cushman (eds) Biodiversity and Ecosystem Processes in Tropical Forests
© Springer-Verlag Berlin Heidelberg 1996

Recently, however, some research and analysis on the roles of individual species and suites of species in maintaining ecosystem processes has been initiated as part of a program sponsored by the Scientific Committee on Problems of the Environment (SCOPE) (Schulze and Mooney 1993). This program explores the ways in which alteration of the species composition of ecosystems affects how ecosystems operate. Specifically, how and to what degree do the key functional processes of ecosystems depend upon the richness of species in them, and how will these properties be changed by the losses of some of their component species? The program was launched by a symposium that was held in Bayreuth, Germany, October 1-4, 1991.

This symposium was followed by 15 workshops, each of which focussed on a major ecosystem, habitat type, or geographical region. One of these workshops[4], the subject of this Volume, was held at Oaxtepec, Morelos, Mexico, December 4-7, 1993. Participants in this workshop examined the role of biodiversity in maintaining ecosystem processes in tropical forests. The processes upon which the workshop focused were primary productivity; secondary productivity; materials processing; provision and maintenance of structure; and some components of ecosystem dynamics, such as resistance to invasion, recovery from disturbance, and ecosystem services. We commissioned eight papers that were written prior to the workshop. A respondent reviewed and commented upon each paper. Workshop participants received the manuscripts and commentaries in advance so that the bulk of the workshop time was devoted to discussions of the papers and the issues they raised. The workshop organizers wrote a synthesis chapter summarizing and integrating the discussions after the workshop. This Volume presents the commissioned chapters, as modified by the workshop discussions, and the synthesis chapter.

Tropical regions are characterized by the relative lack of seasonal changes in temperature, but total annual rainfall and the length and severity of dry seasons varies strikingly with topographic position and latitude. Seasonality of rainfall exerts a strong influence on temporal patterns of primary and secondary production (Janzen and Schoener 1965; Opler et al. 1976; Lieberman 1982; Leighton and Leighton 1983; Bullock and Sólis-Magallanes 1990; Loiselle 1991) and on temporal variation in rates of decomposition (Birch 1958; Jordan 1985; Leigh et al. 1990). Species richness in most taxa of macroorganisms is positively correlated with annual rainfall (Primack 1993) and inversely correlated with the length of the dry season. Both variables are strongly correlated in tropical regions.

---

[4] Persons indicated by an asterisk prepared materials for the workshop but did not attend.

* Seth W. Bigelow, Department of Botany, University of Florida, Gainesville, Florida 32611, USA

* Sandra Brown, Department of Forestry, University of Illinois, 1102 S. Goodwin, Urbana, Illinois 61801 USA

Paul Colinvaux, Smithsonian Tropical Research Institute, Apartado 2072, Balboa, Ancon, Republic of Panama. US mailing address: APO AA 34002-0948.

* C. S. Holling, Department of Zoology, University of Florida, Gainsville, Florida 32611, USA

Moist lowland tropical forests are characterized by both high richness of species in many taxa and complex biotic interactions and linkages. Most tropical plants are animal-pollinated (Bawa 1979, 1990; Bawa and Beach 1981; Baker et al. 1983; Bawa and Hadley 1990), they are fed upon by a wide variety of animals (Dirzo 1987), and they also depend upon animals for dispersal of their seeds (Levey et al. 1981; Howe and Smallwood 1982; Estrada and Fleming 1986). Many biologists have assumed that tropical animals are, on average, more specialized in their diets than their temperate counterparts (Janzen 1973, 1980; Gilbert and Smiley 1978; Beaver 1979), but there are insufficient data on the diets of most tropical organisms to either support or reject this view (Marquis and Braker 1993).

The complex topic of the role of biological diversity on ecosystem functioning can be addressed in many ways and at population, community, and ecosystems levels. An important result of the past three decades of theoretical and empirical research on the causes of patterns in species richness is the demonstration that these patterns are the products of complex interacting forces that vary in relative importance in both time and space (Solbrig 1991). Similarly, the consequences of biodiversity for system-level processes are certain to be equally complex; their elucidation will require analysis of many factors operating at many spatial and temporal scales.

We follow Lawton and Brown (1993) in treating "ecosystem processes," "behavior of ecological systems," and "ecosystem functioning" as equivalent terms. We avoid using "ecosystem function" because of its anthropomorphic implications. Ecosystems process materials and energy, and the efficiency and stability with which they do so is likely to be influenced by biodiversity. Workshop participants explored how and why biodiversity might matter in dynamic processes in tropical forests. By "biodiversity" we mean not only the number of species (species richness) but also the richness of evolutionary lineages, the variety of functional groups of organisms, and the variety of types of ecological communities. Because the number of species in all ecological communities, especially tropical ones, greatly exceeds the number of major ecological processes, many species are involved in each process. We refer to groups of species that participate in a particular process as a *functional group*. A given species may be a member of several functional groups, that is a species may participate in more than one ecosystem process. When more than one species participates in a process we may speak of "functional redundancy," but this term does not imply that all species within the functional group can be substituted for one another with respect to maintaining that process. To the extent that species differ in the quality and quantity of their contributions to some process, the loss of one species will not be compensated for by complementary performance of the others. All trees capture energy by means of photosynthesis, but not all of them grow at the same site or interact with other organisms in the same way.

Functional redundancy makes sense only with reference to some particular process. For example, because many plants on a site carry out photosynthesis using the same mechanism, there may be high functional redundancy in a community with respect to photosynthesis and primary production. However, these plants support species-specific herbivores (Janzen 1973, 1981) and fungi, produce highly distinctive litter (Hobbie 1992), and utilize nutrients from different parts of the soil profile. With respect to these processes, different species in the photosynthesis functional group may exert different influences on other ecosystem processes. Therefore, a species may be a major actor with respect to one process but a minor actor with respect to another.

If the loss of a species results in a large effect on some functional property of an ecosystem, that species may be called a *keystone species*. Keystone species are likely to be found in functional groups with few species or in functional groups having a species whose performance cannot be filled by other group members. For example, a predator that preys selectively upon a competitive dominant may create ecological opportunities for many species that would otherwise be suppressed by the dominant competitor. Traditionally, ecologists have looked for and identified keystone species by their effects on the species richness and composition of the community in which they live. Here we explore the role of taxa that have major consequences for ecosystem processes, such as primary and secondary productivity and nutrient cycling. In this context, a keystone species may or may not significantly change the species composition of its community.

Species within a functional group differ in the magnitude of their contributions, in terms of both their quantity and their quality. For example, if one species of herbivore feeds on plants with heavily defended leaves, whereas another feeds on plants with lightly defended leaves, they may exert different effects on the quantity and quality of leaf litter and, thereby, on humus mineralization rates even if they consume similar quantities of leaves. Similarly, two different frugivores that remove an equal number of seeds of some plant may differ in how well those seeds survive in their digestive tracts and where the frugivore deposits them. One frugivore might be a high-quality disperser; the other a low-quality disperser (Howe and Smallwood 1982). Species within a functional group may also differ in where they carry out these processes. For example, one organism might fix nitrogen on a branch high in the canopy of a forest, another in the soil. Where the fixation occurs influences which other organisms have access to that nitrogen and how the nitrogen moves through the ecosystem.

With respect to any particular process, the number of functional groups is termed the diversity (richness) *of* functional groups. We use the term diversity (richness) *within* functional groups to refer to the number of taxa included in the group. We use richness to mean a simple listing of taxa, such as species or aggregates of them. Diversity is applied to lists that are weighted by some meaningful ecological criterion, such as productivity, biomass, or longevity.

Analyzing the functional significance of biodiversity is a difficult task because the richness of species in most communities is unknown and ecological roles of many described species are poorly understood. In addition, no accepted classifications of functional groups already exist, and no single classification can aggregate organisms appropriately for more than one major ecosystem process. In our analysis of tropical forests we use two major ecosystem processes – energy flow and materials cycling – as the primary basis for establishing functional groups (Lugo and Scatena 1992). We analyze these processes by examining interfaces at which most of the energy or materials are exchanged. At each of these interfaces there is a discontinuity of resource availability that is used by groups of species as an energy or nutrient source (Table 1.1).

Because energy is consumed and not recycled, flows along most energy pathways are unidirectional, and much energy is lost as heat at each transfer. Flow of materials at interfaces is typically bidirectional but transfer rates at a particular interface are often unequal. Changes in relative rates of transfer of materials at interfaces may trigger major changes in ecosystem functioning, with the result that species that influence transfer rates are likely to be keystone species.

The amount of energy flow at different interfaces may be a poor indicator of the significance of an interface for the ecosystem processes that are affected. For example, the transport of a small amount of energy by a pollinator may catalyze large investments by plants in fruit and seed production, with consequent effects on population sizes and dynamics of frugivores, and, on longer time frames, recruitment of the plants. Similarly, the

Table 1.1. Ecological interfaces where large amounts of energy or materials are exchanged. (Adapted from Silver et al. Chap.4 this Vol.)

Interface	Exchange processes
Atmospheric - organism	Photosynthesis, transpiration, capture, and retention of nutrients, release of CO_2, CH_4, NO, N_2O, SO_2, etc.
Plant-soil	Nutrient capture and recapture, nutrient storage, nutrient transfer
Biotic	
Plant-herbivore	Consumption of plant tissues, storage and translocation of nutrients
Animal-animal	Predation, parasitism, mutualism
Detritus-detritivore	Decomposition
Terrestrial-hydrologic	Recapture of nutrients from groundwater, loss of nutrients to groundwater

quantities of nutrients transferred may be an inadequate measure of the importance of an interface for ecosystem processes. Mobilization or immobilization of modest amounts of nutrients may also trigger large responses on the parts of organisms. Such magnifier effects are explored in the chapters of this book.

In Chapter 2, Joseph Wright[5] examines the role of biological diversity in generating and maintaining primary productivity and the variety of ways in which the carbon fixed during photosynthesis is allocated to different tissues (wood, roots, leaves, sap, flowers, and fruit) in tropical forests. In Chapter 3, Michael Huston[6] and Larry Gilbert[7] use the complex packaging of primary production as input for an analysis of the ease with which the different forms of carbon can be utilized and the location of the major rate-controlling processes along each of these energy flow pathways. Whendee L. Silver[8], Sandra Brown, and Ariel Lugo explore the role of biodiversity in processing of nutrients in tropical forests, paying special attention to sites of nutrient transfers and the processes most important for conserving nutrients in the system (Chap. 4). The role of fungal and bacterial diversity on forest functioning is explored by Jean Lodge[9], David Hawksworth[10], and Barbara Ritchie in Chapter 5. Taken together, these four chapters provide a basic overview of the functional groups we found useful for analyzing productivity and materials processing by tropical forests.

The remaining chapters examine special aspects of tropical forest processes. In forested environments trees provide most of the physical structure within which most other organisms live, as well as being the source of most of the primary production upon which all other organisms depend. Therefore, the role of the diversity of plant life-forms warrants and receives extensive treatment by John Ewel[11] and Seth Bigelowin Chapter 6. Recognizing that tropical forests are continually subjected to various perturbations that may threaten the maintainance of key processes, in Chapter 7 Julie Denslow[12] explores the potential role of biological diversity, at levels of species and functional groups, on resistance to and recovery from disturbance. One particular form of disturbance – invasions of exotic species – is so

[5] S. Joseph Wright, Smithsonian Tropical Research Institute, Apartado 2072, Balboa, Ancon, Republic of Panama. USA mailing address: Unit 0948. APO AA 34002-0948

[6] Michael Huston, Environmental Division, Oak Ridge National Laboratory, P. O. Box 2008, Oak Ridge, Tennessee 37831-6038 USA

[7] Larry Gilbert, Department of Zoology, University of Texas, Austin, Texas 78712, USA

[8] Whendee L. Silver, Department of Biology, Boston University, Boston MA 02215

[9] D. Jean Lodge, USDA, Forest Service, Forest Products Laboratory, P. O. Box 1377, Luquillo, Puerto Rico 00773-1377

[10] David L. Hawksworth, International Mycological Institute, Bakeham Lane, Egham, Surrey, TW20 9TY, UK

[11] John J. Ewel, Department of Botany, University of Florida, Gainesville, Florida 32611, USA

[12] Julie Denslow, Department of Botany, Louisiana State University, Baton Rouge, Louisiana 70803, USA

important today that we devote Chapter 8 to Marcel Rejmanek's[13] examination of the importance of biodiversity for ecosystem responses to invading species. Biodiversity and functioning of tropical managed ecosystems are explored by Alison Power[14] and Alexander Flecker[15] in Chapter 9.

Flowing from these chapters, and the workshop discussions that helped mold and modify them, are trends and generalizations that cut across the foci of individual contributions. We gather together and synthesize these issues in Chapter 10. For example, how do ecosystem processes, and the functional groups that carry them out in tropical forests, change across gradients of rainfall, soil, fertility, and altitude? Chapter 10 also considers the role of biological diversity for the responses of tropical forests to a range of threats, some of which are increasing in frequency and severity today. The analyses in the first nine chapters highlight many gaps in knowledge which impede our predictions of forest responses, our abilities to continue to use forests as sources of knowledge and insights for understanding complex systems, restoring damaged and degraded forests, designing appropriate management strategies, and extracting commercially valuable products from tropical forests on a sustained basis. Suggestions for high priority research topics that should greatly improve our understanding of the role of biological diversity in tropical forest processes are an important part of that final chapter. The need for more complete knowledge is vast, but resources and time are in limited supply. If we have made a contribution that can help target future research in a more productive way, the workshop will have achieved one of its basic goals.

---

[13] Marcel Rejmánek, Department of Botany, University of California, Davis, Davis, California 95616, USA

[14] Alison Power, Section of Ecology & Systematics, Cornell University,

[15] Alexander S. Flecker, Section of Ecology and Systematics and Center for the Environment, Cornell University, Ithaca, New York 14853, USA

References

Baker HW, Bawa KS, Frankie GW, Opler PA (1983) Reproductive biology of plants in tropical forests. In: Golley FB (ed) Tropical rain forest ecosystems: structure and function. Elsevier, New York, pp 183-205

Bawa KS (1979) Breeding systems of trees in a tropical wet forest. N Z J Bot 17:521-524

Bawa KS (1990) Plant-pollination interactions in tropical lowland rain forest. Annu Rev Ecol Syst 21:254-274

Bawa KS, Beach JH (1981) Evolution of sexual systems in flowering plants. Ann Mo Bot Gard 68:254-274

Bawa KS, Hadley M (eds) (1990) Reproductive ecology of tropical forest plants. UNESCO, Parthenon

Beaver RA (1979) Host specificity of temperate and tropical animals. Nature 281:139-141

Birch HF (1958) The effect of soil drying on humus decomposition and nitrogen availability. Plant Soil 10:9-13

Bullock SH, Sólis-Magallanes JA (1990) Phenology of canopy trees of a tropical deciduous forest in Mexico. Biotropica 22:22-35

Dirzo R (1987) Estudios sobre interacciones herbívoro-planta en Los Tuxtlas, Veracruz. Rev Biol Trop 35:119-131

Estrada LE, Fleming TH (eds) (1986) Frugivores and seed dispersal. Junk, Dordrecht, 392 pp

Food and Agricultural Organization (1993) Tropical forest resources assessment project. FAO, Rome

Gentry AH (1982) Patterns of neotropical species diversity. Evol Biol 15:1-84

Gilbert LE, Smiley JT (1978) Determinants of local diversity in phytophagous insects: host specialists in tropical environments. In: Mound LA, Waloff N (eds) Diversity of insect faunas. Symp R Entomol Soc Lond.9:89-105

Green GM, Sussman RW (1990) Deforestation history of the eastern rain forests of Madagascar from satellite images. Science 248:212-215

Hobbie SE (1992) Effects of plant species on nutrient cycling. Trends Ecol Evol 7:336-339

Howe HF, Smallwood J (1982) Ecology of seed dispersal. Annu Rev Ecol Syst 13:201-228

Janzen DH (1973) Comments on host-specificity of tropical herbivores and its relevance to species richness. In: Heywood V (ed) Taxonomy and ecology. Academic Press, New York, pp 210-211

Janzen DH (1980) Specificity of seed-attacking beetles in a Costa Rican deciduous forest. J Ecol 68:929-952

Janzen DH (1981) Pattern of herbivory in a tropical deciduous forest. Biotropica 13:271-282

Janzen DH, Schoener TW (1968) Differences in insect abundance between wetter and drier sites during a tropical dry season. Ecology 49:96-110

Jordan CF (1985) Nutrient cycling in tropical forest ecosystems. Wiley, New York

Lawton JH, Brown VK (1993) Redundancy in ecosystems. In: Schulze E -D, Mooney HA (eds) Biodiversity and ecosystem function. Springer, Berlin Heidelberg New York, pp 255-270

Leigh EG, Rand AS, Windsor DM (eds) (1990) Ecología de un bosque tropical: ciclos estacionales y cambios a largo plazo. Smith Trop Res Inst, Panama, 546 PP

Leighton M, Leighton DR (1983) Vertebrate responses to fruiting seasonality within a Bornean rain forest. In: Sutton SL, Whitmore TC, Chadwick AC (eds) Tropical rain forest: ecology and management, Blackwell Scientific, London, pp. 181-196

Levey DJ, Moermond TC, Denslow JS (1993) Frugivory: an overview. In: McDade LA, Bawa KS, Hespenheide HA, Hartshorn GS (eds) La Selva. Ecology and natural history of a neotropical rain forest. Univ Chicago Press, Chicago, pp 282-294

Lieberman D (1982) Seasonality and phenology in a dry tropical forest in Ghana. J Ecol 70:791-806

Loiselle BA (1991) Temporal variation in birds and fruits along an elevational gradient in Costa Rica. Ecology 72:180-193

Lugo AE, Scatena FN (1992) Epiphytes and climate change research in the Caribbean: a proposal. Selbyana 134:123-130

Marquis RJ, Braker HE (1993) Plant-herbivore interactions: diversity, specificity, and impact. In: McDade LA, Bawa KS, Hespenheide HA, Hartshorn GS (eds) La Selva. Ecology and natural history of a neotropical rain forest. Univ Chicago Press, Chicago, pp 261-279

Opler PA, Frankie GW, Baker HG (1976) Rainfall as a factor in the release, timing, and synchronization of anthesis by tropical trees and shrubs. J Biogeogr 3:231-236

Primack RB (1993) Essentials of conservation biology Sinauer, Sunderland, MA

Schulze E-D, Mooney HA (eds) 1993. Biodiversity and ecosystem function. Springer, Berlin Heidelberg New York

Skole D, Tucker D (1993) Tropical deforestation and habitat fragmentation in the Amazon: satellite data from 1978-1988. Science 260:1905-1910

Solbrig OT (ed) (1991) From genes to ecosystems: a research agenda for biodiversity. IUBS, Cambridge, MA

Whitmore TC, Sayer JA (eds) (1992) Tropical deforestation and species extinction. Chapman & Hall, London

Wilson EO (1992) The diversity of life. Belknap, Cambridge

2 Plant Species Diversity and Ecosystem Functioning in Tropical Forests

S. Joseph Wright[1]

2.1 Introduction

Ecosystem processes emerge from the capture, transfer, and loss of energy and nutrients by biological species. It has been clearly established that both the presence of particular species and also the total number of species involved can influence emergent ecosystem processes. Keystone species such as predators that prey selectively on competitive dominants or plants that fruit during seasons of scarcity can have profound effects on entire ecosystems and have been identified in a wide range of habitats from the rocky intertidal to tropical forests (Paine 1966; Terborgh 1986). In contrast, evidence that the total number of species present influences ecosystem functioning has been limited to laboratory microcosms (Naeem et al 1994) and to relatively species-poor grasslands (McNaughton 1993; Tilman and Downing 1994). In this chapter, the question is asked whether the total number of plant species present affects some components of ecosystem functioning in tropical forests. Tropical forests include the most floristically diverse habitats on the planet. Although it is well established that species richness influences processes in species-poor ecosystems, the effect of an increase from perhaps 50 tree species per hectare in a dry tropical forest to more than 300 tree species per hectare in a wetter forest is uncertain. Uncertainty also remains over when, where, and how species composition affects ecosystem functioning.

This chapter addresses the relationship between plant species diversity and ecosystem productivity and stability. In the introductory chapter to this volume, Orians et al. provide a more complete catalog of ecosystem processes. Here consideration is limited to productivity and stability because the relationship between tropical forest plant species diversity and other ecosystem properties has been reviewed elsewhere (Vitousek and Hooper 1993) and is explored in subsequent chapters in this book. To avoid

[1] Smithsonian Tropical Research Institute, Apartado 2072, Balboa, Ancon, Republic of Panama.

Ecological Studies, Vol. 122
Orians, Dirzo and Cushman (eds) Biodiversity and Ecosystem Processes in Tropical Forests
© Springer-Verlag Berlin Heidelberg 1996

Table 2.1. Definitions of terms associated with ecosystem stability. (Pimm 1984)

	BIODIVERSITY
Species richness	The number of species in a system
Evenness	The variance of species abundances. A numerically dominant species increases this variance and reduces evenness
Connectance	The number of interspecific interactions that actually occur divided by the potential number
Interaction strength	The mean effect of one species' density on the growth rate of other species
	STABILITY
Stable	An ecosystem is stable if and only if all variables return to equilibrium values following a perturbation away from equilibrium. Variables of interest include species composition and individual species abundances.
Resistance	The degree to which a variable is changed by a perturbation
Resilience	The rate at which a variable returns to equilibrium following a perturbation
Variability	The variance of population densities over time

ambiguity, the definitions of diversity, stability, and associated terms (Table 2.1) are adopted from Pimm (1984), where they are more fully developed. Mechanisms that relate species diversity and ecosystem processes are explored, then productivity with different plant species diversities is compared. Finally the stability of paleoforests followed through time after major perturbations is evaluated for modern tropical forests.

2.2 The Dependence of Ecosystem Processes on Species Diversity

The dependence of ecosystem processes on species diversity varies with species' ecological breadths. Every plant species performs best over some subset of the environmental conditions found in nature, and performance declines away from this set of environmental conditions (Austin and Smith 1989). This concept is represented by species performance functions that describe performance along environmental gradients (Fig. 2.1A). If each species performed well over a narrow set of environmental conditions, species-rich communities could have close species packing with little overlap among species (May and MacArthur 1972). If, in addition, the environmental factors that determined species performance varied over space or time, then the addition or deletion of a species would alter ecosystem functioning (McNaughton 1993). In the extreme case presented in Fig. 2.1A and B, the addition of each species changes ecosystem functioning (Fig. 2.1C). Alternatively, if some species perform well over a broad ~et of environmental conditions (Fig. 2.1D), then there would necessarily be ⌐ overlap among species and functional redundancy in species-rich com-

broad overlap among species and functional redundancy in species-rich communities. In the extreme case presented in Fig. 2.1D and E, overlap and functional redundancy are so great that species diversity and ecosystem functioning are largely independent (Fig. 2.1C). Both views of tropical forest plant communities have adherents.

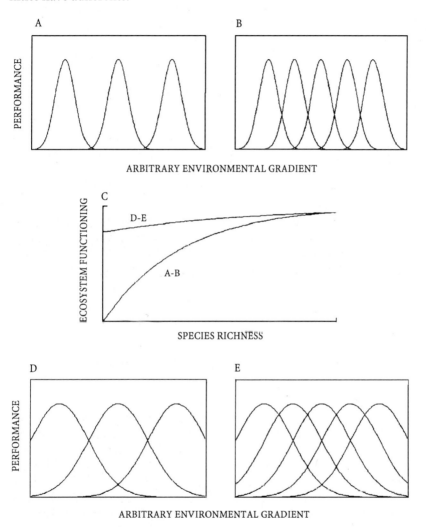

Fig. 2.1A-E Two conceptual models of plant communities and their consequences for the relation between plant species richness and ecosystem functioning (**C**). A and B represent a low and a high richness community, respectively, in which each species performs well over a narrow range of environmental conditions and there is little overlap among species. **D** and **E** represent a low- and a high-richness community, respectively, in which each species performs well over a broad range of environmental conditions and there is substantial overlap and functional redundancy among species. The *curves* represent fundamental niches or performance in monoculture. Austin and Smith (1989) discuss the nested or included nature of such performance curves in plant communities

One extreme view is that each tropical forest species is competitively superior in some subset of environmental conditions. Proponents of this view argue that the great age of tropical forests and the relative constancy of perhumid tropical climates have allowed a slow accumulation of new species and competitive coevolution of species, leading to extreme specialization and extremely tight species packing, as in Fig. 2.1B (Ashton 1969, 1993). Another equally extreme view is that plant species are ecologically equivalent, as in Fig. 2.1E, and that high diversity is maintained by disturbance and chance alone (Hubbell and Foster 1986a). Reality probably lies somewhere between these extremes.

Investigations of the ecological differences among tropical forest plants have had mixed results. Specialization to narrow ranges of edaphic conditions and to narrow ranges of light gradients associated with treefall gaps have been predicted to be important factors permitting coexistence of tropical forest trees (Ashton 1969 1993; Ricklefs 1977; Denslow 1980). A high degree of specialization to narrow ranges of gap size was indeed found among three heliophilic tree species in Panama (Brokaw 1987); but growth and survivorship of the great majority of tree species were indifferent to gaps in the same forest (Welden et al. 1991). Recent experimental studies have also failed to find species differences in response to light gradients associated with treefall gaps (Denslow et al. 1990). The degree of specialization to edaphic conditions also varies among species. Soil drainage restricted the distributions of several tree species in Panama and also in Costa Rica, but the great majority of tree species at both sites were widely distributed from seasonally waterlogged to well-drained soils (Lieberman et al. 1985; Hubbell and Foster 1986b). Likewise, soil fertility influenced the distributions of several tree species in Borneo. Once again, however, most species grew over a broad range of soil fertilities (Baillie et al. 1987). Although each of these studies identified species with restricted ecological tolerances, many species were indifferent to the environmental gradients. These indifferent species clearly have broadly overlapping ecological niches.

These and other studies have been limited to small parts of species' ranges. Added perspective is gained by considering the full range of habitats where a species is common. Three Neotropical examples follow. In central Panama, *Prioria copaifera* Griseb. is common on well-drained upland soils and also forms monodominant stands in freshwater swamps. *Tabebuia rosea* (Bertol.) DC. and *Bombacopsis quinata* (Jacq.) Dugand are characteristic of tropical dry and moist forests in Panama and are also common on sand and rocks, respectively, along Panama's dry Pacific beaches. *Ficus insipida* Willd. is a large, abundant tree of early successional forests on well-drained soils in Central America and also of seasonally flooded soils in the Varzea forests of the Amazon. The conclusion that some tropical forest plants perform well over a range of environmental conditions that is broad relative to the range of conditions occupied by tropical forests is inescapable.

The possibilities outlined in Fig 2.1 can now be evaluated. It is indisputable that many tropical forest plant species have narrow ecological requirements (Gentry 1986), and ecological differences are likely among such species. It is equally clear, however, although perhaps not yet generally acknowledged, that many tropical forest plant species have broad ecological tolerances and broadly overlapping performance functions. A high degree of overlap and functional redundancy is inevitable when species with broad ecological tolerances are present and species-richness reaches the high levels commonly observed in tropical rain forests. For these reasons, ecosystem productivity can be predicted to reach an asymptote with respect to plant species richness at levels of plant species richness well below those commonly observed in tropical forests.

2.3 Plant Species Richness in Tropical Forests

Two comments about the plant species richness of tropical forests are in order before this prediction is tested. Tropical forest plant species diversity will first be evaluated using the definitions of diversity proposed by Pimm (1984), and second, correlates of plant species diversity that might confound comparative analyses of ecosystem functioning in tropical forests will be considered.

Pimm (1984) proposed four definitions of diversity (Table 2.1). The first, species richness, is simply the number of species present. Only one point estimate of the species richness of vascular plants in a tropical rain forest setting exists. Gentry and Dodson (1987) found 365, 169, and 173 vascular plant species in 1000-m^2 samples from three Ecuadorian rain forests, levels that far exceed those observed in any other ecosystem. The species richness of trees alone can also be remarkably high with more than 300 large tree species recorded from 1-ha plots in western Amazonia and Borneo (Gentry 1988a; Ashton 1993; Valencia et al. 1994).

The three remaining diversity measures proposed by Pimm (1984) can be evaluated only for trees, the data from tropical forests being inadequate for other plant life forms. Pimm (1984) defined evenness as the variance of species abundances. Numerically dominant species that increase this variance and decrease evenness are rare in tropical forests, and evenness is often exceptionally high (Gentry 1988b). For example, in Amazonian Ecuador, ha plots included 300 and 307 species among their 606 and 696 individual large trees, respectively (Gentry 1988a; Valencia et al. 1994). Pimm (1984) defined interaction strength as the mean effect of one species' density on the growth rate of other species. Interaction strength will be large and broadly similar across favorable habitats, including tropical forests with closed plant canopies and intense competition for space and light. Finally, Pimm (1984) defined connectance as the number of interspecific interactions that actually occur divided by the potential number. For sessile plants, interactions are largely limited to near neighbors and connectance can be measured by the species richness of near neighbors. For tropical forest trees, the species composition of near neighbors, approximates a random draw from the local species pool (Hubbell and Foster 1986a), and connectance is large. By all definitions, tropical forests have exceptionally high plant diversities.

The species diversity of tropical forest plants is not uniformly large, however. Diversity varies with the abiotic environment, history, and biogeography. This variation provides the raw material for the comparisons to be developed here, but caution is required. In particular, differences in abiotic environments may confound apparent relationships between diversity and ecosystem functioning whenever the abiotic environment has direct effects on ecosystem functioning.

Two examples will illustrate the problem. Rainfall is a critical abiotic variable in tropical forests. The species richness of trees increases five- and sixfold as annual rainfall increases from 1000 to 4000 mm in the Neotropics and from 750 to 1750 mm in Ghana (Hall and Swaine 1981; Gentry 1988a). Rainfall also has direct effects on ecosystem functioning. Drought limits primary production in tropical forests (Brown and Lugo 1982; Medina and Klinge 1983), decreases resistance to some types of perturbations such as fire (Woods 1989; Hart et al. 1995), and decreases plant growth rates and ecosystem resilience after perturbations. Drought is also more frequent in drier tropical forests (unpublished analyses of rainfall seasonality and supraannual variability). As a consequence, productivity and stability of productivity vary along tropical forest rainfall gradients for reasons unrelated to differences in plant diversity.

Soil fertility also affects the species diversity of tropical forest plants (Wright 1992). An inverse relation has been reported between tree species diversity and soil fertility in Costa Rica and Ghana (Huston 1980, 1994). However, in both instances a strong positive relationship between diversity and rainfall confounds causality (Holdridge et al. 1971; Hall and Swaine 1981). In contrast, the diversity of the Dipterocarpaceae is greatest at intermediate soil fertilities in Borneo (Ashton 1977); and diversity increases with soil fertility for understory herbs and shrubs in the Neotropics; for trees in the families Rubiaceae, Meliaceae, and Euphorbiaceae in Borneo; and for trees in the Neotropics (Gentry and Emmons 1987; Gentry 1988a; Ashton 1988). As with rainfall, soil fertility has direct effects on ecosystem processes such as productivity. For these reasons, it is essential to control abiotic variables in comparative studies of the relationship between plant species diversity and ecosystem functioning.

2.4 The Primary Productivity of Tropical Forests

Productivity for species-rich and species-poor forests was compared to test the hypothesis that primary productivity reaches an asymptote at levels of plant species richness well below those commonly observed in tropical forests. The analyses are restricted to the production of fine litter (leaves, fruit, flowers and small branches) because there are too few estimates of the other components of productivity for tropical forests.

To provide perspective on the following analyses of fine litter, the magnitude of all components of tropical forest productivity are briefly described. Net primary production (NPP) includes fine litter, wood, roots, pollen, nectar, and plant exudates including resins and root exudates.

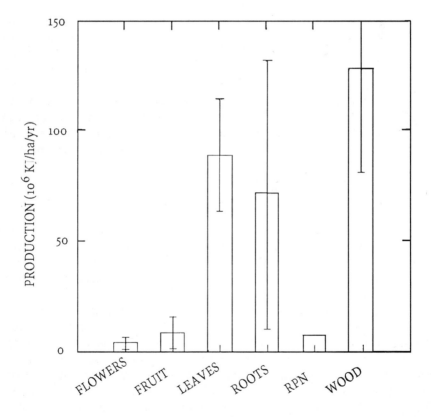

Fig. 2.2. The components of net primary production. *RPN* represents resin, pollen, and nectar combined. *Histograms* and *error bars* represent means and one standard deviation, respectively. Sources are nectar, pollen and resin (Roubik 1993); roots (Kira 1978; Jordan and Escalante 1980; Cuevas and Medina 1988; Cavelier and Wright (1995); wood (Jordan 1983); flowers, fruits, and leaves (Wright and Cornejo, in prep.). Differences in methods contribute to the large standard deviation for roots. Conversion factors given by Golley (1969) were used to convert mass production estimates to kJ ha^{-1} yr^{-1}

Gross primary production (GPP) also includes plant respiration, which accounted for 67, 75 and 87% of GPP in the three available tropical forest estimates (Muller and Nielsen 1965; Odum 1970; Kira 1978). NPP ultimately supports animal respiration, and the division of energy among the components of NPP is critical to forest heterotrophs (see Huston and Gilbert, Chap. 3, this Vol.). Wood, leaves, and roots form the largest components of NPP (Fig. 2.2). Estimates for leaves, flowers, and fruit are from litterfall and are uncorrected for consumption by arboreal animals. The three available estimates for root production are also low because large roots (usually > 2 mm diameter) are excluded. The single available estimate for resin, pollen, and nectar production for a tropical forest deserves comment. Roubik (1993) calculated the energetic demands of resin, pollen, and nectar consumers for Barro Colorado Island, Panama. He concluded that resin, pollen, and nectar consumption accounted for 3% of above-ground NPP and exceeded leaf and fruit consumption by vertebrates and leaf consumption by phytophagous insects. There are no estimates of tropical forest root exudate production.

In contrast, there are numerous estimates of leaf litter production. The leaf litter that falls to the forest floor accounts for 25 to 60% of NPP in tropical forests, and 50% of the variation in this proportion is explained by the ratio of mean temperature to annual precipitation (Brown and Lugo 1982). Therefore leaf litterfall was comparefd for closed canopy forests with similar ratios of mean temperature to annual precipitation but very different plant species diversities. Because many of the original sources did not separate leaf litterfall from other fine litterfall (flowers, fruit, and small branches), the analyses were repeated for all fine litter.

The first comparison contrasted forests where plant species diversity was high with forests where one or a few tree species dominated the canopy. Such forests are found throughout the tropics although their areal extent is often small (Connell and Lowman 1989; Hart et al. 1989). Production, temperature and rainfall data were taken from Proctor (1984) and augmented by interpolation from standard meteorological references. Analyses of covariance were performed in which the covariate was the ratio of mean temperature to annual precipitation, the dependent variable was leaf or total fine litterfall, and the grouping variable was plant species diversity. Both fine litter- and leaf litterfall were highly variable; however, there were no differences between diverse forests and those dominated by a few tree species (Fig. 2.3, p > 0.4).

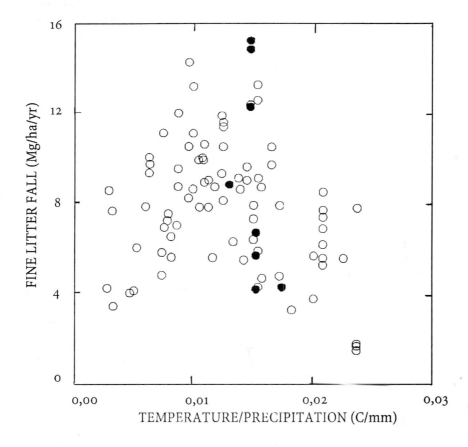

Fig. 2.3. The relationship between total fine litterfall (Mg ha^{-1} yr^{-1}) and the ratio of mean annual temperature to annual rainfall (°C mm^{-1}). *Open* and *solid* circles represent tropical forests with high diversity versus tropical forests with heavy dominance by one or a few tree species, respectively. Brown and Lugo (1982) discuss this relationship in greater detail

The second comparison contrasted mainland forests where plant species diversity was high with oceanic island forests characterized by low species diversity (MacArthur and Wilson 1967). Puerto Rico and Hawaii are the only tropical islands with published production data (Proctor 1984) so mainland sites were limited to similar latitudes (15 ≤ latitude ≤ 23). Fine litterfall was unaffected by the ratio of mean temperature to annual precipitation over this limited latitudinal range (p > 0.4), and a one-way analysis of variance was used to compare island and mainland forests. Fine litterfall was significantly lower for the islands (Fig. 2.4, p = 0.015). Additional data are needed to determine whether low litter production is a peculiar property of Puerto Rican and Hawaiian forests or a characteristic of depauperate island forests.

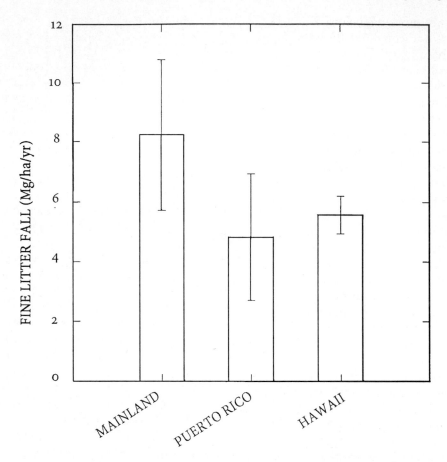

Fig. 2.4. Fine litterfall for mainland tropical forests characterized by high plant species richness and tropical forests from Puerto Rico and Hawaii. Mainland forests are from latitudes similar to Puerto Rico and Hawaii. *Histograms* and *error bars* represent means and one standard deviation, respectively

There has been one experimental study of the relation between plant species diversity and primary production in a tropical forest setting. Ewel and associates (Brown and Ewel 1987; Berish and Ewel 1988) followed forest succession for 5 years after clearing and burning forest plots at Turrialba, Costa Rica. Their four treatments included natural succession, an enriched succession in which seeds of introduced species were added to the natural succession, an imitation of succession in which each natural colonist was removed and replaced with an introduced species of the same life form, and a temporal sequence of monocultures with the species grown in monoculture chosen to match the predominant life form in the first three treatments.

Above- and belowground measurements of primary production were made throughout the experiment, but have not yet been fully published. Preliminary results are intriguing. The productive capacity of any vegetation depends on the total leaf area per unit soil area (leaf area index, LAI). After 47 months, when trees first dominated all four treatments, LAI varied only between 3.33 and 3.44 m^2 m^{-2} even though the mean number of plant species per experimental plot was 104, 107, 133 and 1 for natural, imitation and enriched succession, and the monoculture, respectively (Brown and Ewel 1987). Primary production is unlikely to have varied substantially at this stage of succession. Although there were early treatment differences in root development, after 5 years the monoculture and natural succession treatments supported similar biomasses of both fine (< 2 mm diameter) and coarse roots (> 2 mm; Berish and Ewel 1988). Plant species diversity and primary production were largely unrelated during the early stages of tropical forest succession.

Comparisons of species-poor tropical forest plantations and species-rich native tropical forests are also suggestive. A broad survey found similar wood production rates in plantations and native forests (Jordan 1983). A more detailed comparison of plantations and natural successional forests of the same age found higher aboveground NPP but lower root biomass in plantations (Lugo 1992). In conclusion, similar litter production by high and low diversity mainland forests (Fig. 2.3), similar wood production in plantations and native forests, and similar LAI and root biomasses in the only relevant experimental study, all indicate that plant species diversity has little influence on primary production in tropical forests.

There are two reasons to exist which could have anticipated this result. The productive capacity of a forest depends on the photosynthetic efficiency of its leaf tissue and on LAI. Photosynthetic efficiencies vary little among the C3 plants that dominate mature tropical forests (Bjorkman and Demmig 1987). For this reason, increased species diversity is likely to increase forest productivity only if LAI is increased. A new species might increase effective LAI by presenting leaves in previously unoccupied space or by presenting leaves when other plants are deciduous.

Neither possibility is likely in mature closed canopy tropical forest. A new leaf in a previously unoccupied space shades lower leaves. As shade becomes more intense, respiration eventually exceeds photosynthesis and the shaded leaves are abscised (Addicott 1982). Even the addition of new life forms is unlikely to increase LAI. For example, lianas displace rather than augment the leaf area of their host trees (Ogawa et al. 1965; Kira and Ogawa 1971). Epiphytes that rest on and shade opaque branches may be an exception and may increase LAI (Ewel and Bigelow, Chap.6, this Vol). Nonetheless, LAI varies little among mature tropical forests that share similar physical environments (Medina and Klinge 1983).

The potential for different temporal presentations of leaves to increase effective LAI averaged over all seasons is also limited. Most humid tropical forests experience one or two annual dry seasons (Richards 1952). Plant responses to the drier seasons fall along a continuum associated with access to soil water. Plants with shallow roots and plants in relatively xeric micro-sites may experience water stress and become deciduous (e.g., Reich and Borchert 1984). In contrast, plants with deeper roots and plants in relatively mesic microsites may escape water stress (Wright and Cornejo 1990; Wright and Van Schaik 1994). Plants in the latter group often produce new leaves as the dry season begins, and flower and fill fruit as the dry season progresses. This may be a response to high dry-season irradiance associated with reduced cloud cover and reduced shading by deciduous neighbors (Van Schaik et al. 1993; Wright and Van Schaik 1994). This continuum of leafing responses is unlikely to increase effective LAI averaged over the year because most deeply rooted plants are evergreen and contribute to LAI year-round. Only the addition of wet-season deciduous species would complement dry-season deciduous species to increase effective LAI. Wet-season deciduous species are exceedingly rare (Janzen 1972), and seasonal tropical forests consist almost entirely of dry-season deciduous and ever-green species (Richards 1952). In conclusion, new species that present leaves in novel positions or at novel times are unlikely to increase LAI in mature stands of tropical forest. Therefore, it is not surprising that NPP and plant diversity are uncoupled in mature tropical forests.

It remains to consider the potential role of species richness during primary and secondary succession and during forest regeneration after treefalls, blowdowns, and other disturbances that create gaps in mature, closed canopy forests. Both forest succession and gap regeneration occur rapidly in the tropics as colonizing, heliophillic species quickly reestablish LAI (Denslow 1987; Brown and Lugo 1990). Large numbers of species are involved in most tropical forests. These include herbs, shrubs, and trees that often produce copious numbers of small seeds that persist in the soil, germinate in high light, grow rapidly, and survive poorly if at all after a closed canopy is reestablished (Denslow 1987; Smith 1987; Garwood 1989). If all such species were eliminated, then averaged over time forest productivity would undoubtedly decline. However, the experiment conducted by Ewel and associates and reviewed earlier indicates that a radical difference in the number of colonizing species has no effect on LAI or on root biomass after 5 years of secondary succession (Brown and Ewel 1987; Berish and Ewel 1988). Again, it is unlikely that species richness will influence forest productivity over the ecologically relevant range of species richnesses observed in tropical forests.

2.5 The Stability of Tropical Forests

Early ecologists were impressed with the apparent stability of tropical forests, but these impressions were based on few data. For example, in an influential paper, Allee (1926) concluded that the constancy of the forest interior environment rivaled the capabilities of growth chambers of the era. His data consisted of measurements from a single season and site. Likewise, Elton (1958) apparently based the conclusion that pest outbreaks were unusually rare in tropical forests on discussions with three tropical foresters. Remarkably, the notion that population densities are highly constant in tropical forests became fixed in the ecological literature without the benefit of a single demographic study of a tropical forest organism. Subsequent studies have contradicted many of these early impressions of tropical forest stability.

Strong population density fluctuations characterize many tropical forest organisms. Population density fluctuations of insects in tropical forests and the north temperate zone are comparable (Wolda 1978, 1988). Tropical forest mammals undergo similar strong population density fluctuations (Foster 1982; Glanz 1982; Giacalone et al. 1990), and among lizards, the species with the greatest population density fluctuations yet observed is from a tropical forest (Andrews and Wright 1994). Remarkable population density fluctuations have also been observed among tropical forest plants. In just 3 years, population densities changed by more than 10% for 40% of the shrub and tree species encountered in a 50-ha plot of mature forest in Panama (Hubbell and Foster 1990). Population outbreaks of plant pests are an extreme form of population variability. A rapid decline in population density of one formerly common tree species was caused by an outbreak of a disease-causing pathogen (Gilbert et al. 1994). There is also increasing evidence that population outbreaks of herbivorous insects are not as rare in tropical forests as once believed (Wolda and Foster 1978; Janzen 1981, 1984, 1985; Wong et al. 1990). Clearly, tropical forests are not particularly stable when judged by the criterion of population variability.

Other early impressions of tropical forest stability may also have to be discarded as more data become available. In particular, tropical forests may not be particularly resistant to invasion by exotic species. Africanized bees, a familiar exotic, have now invaded virtually all Neotropical forests (Roubik 1991). *Drosophila malerkotliana*, a lesser-known native of India, invaded Neotropical forests in the 1970s and has become common and widespread (Bock 1980; Sevenstar 1991). Indigenous peoples may also have been important agents of introduction. As an example, the fruit tree *Chrysophyllum cainito* L. was native to the West Indies but is now naturalized and is a regular component of tropical dry and moist forests in Central America, southern Mexico, and northern South America (Blackwell 1968). Rejmanek evaluates other tropical plant invasions (Chap.8, this Vol.).

These examples suggest that tropical forest may be less resistant to invasion than previously thought.

Direct measures of other aspects of stability such as resilience, resistance, and stability itself are much rarer. No controlled experiments to study tropical forest stablility have been performed, but observations suggest that resistance varies with the type of perturbation. Resistance to hurricane damage is high for species-rich tropical forests and also for species-poor temperate zone forests that are regularly exposed to hurricanes (Brokaw and Walker 1991; Gresham et al. 1991). In contrast, fire may be a truly catastrophic event in tropical forests (Woods 1989). The earlier impression that lowland rain forest was generally immune to fire (Richards 1952) has been overturned by the discovery of widespread charcoal deposits many of which, in the Neotropics, predate human occupation (Sanford et al. 1985). The severe droughts that permit rain forest fires may occur as frequently as once every two or three tree generations both in Borneo, the central Amazon and the Congo Basin (Woods 1989; Hart et al. 1995; Piperno and Becker, 1995). These observations suggest that resistance to perturbation is not particularly well developed in diverse tropical forests.

Natural experiments address tropical forest stability in response to climate change and agriculture. Tropical climates during glacial periods were cooler and, in some places, drier than are modern tropical climates (Hamilton 1982; Piperno et al. 1991a,b). Not surprisingly, the geographic ranges of tropical plants changed in response to Pleistocene climate change. Recent pollen analyses indicate that these movements brought together unfamiliar combinations of tree taxa. In central Panama, for example, *Quercus* and other tree taxa which are now largely limited to elevations above 1000 m coexisted at intermediate elevations (500 to 650 m) with a variety of tree taxa that are now largely limited to still lower elevations (Bush and Colinvaux 1990; Bush et al. 1992). Unfamiliar species mixtures also prevailed in North America as trees reinvaded after glaciation (Prentice et al. 1991). The appearance of forests with novel species compositions in both the species-rich tropics and the species-poor temperate zone again suggests that modern tropical forests are not particularly stable.

A natural experiment is provided by the rise of agriculture among the indigenous peoples of the Neotropics, depopulation following the Spanish Contact, the widespread collapse of agriculture, and forest recovery beginning 500 years B.P. Several pollen and phytolith records now record these events (Leyden 1987; Jones 1991; Piperno et al. 1991a,b; Bush and Colinvaux 1994). Extremely rapid recovery of arboreal taxa characterizes each of these records in the decades immediately following Contact; however, the taxonomic match between the original forest removed by indigenous peoples and the post-Contact recovering forest is not good even after 400 years. Continued low-level human activity and/or climate change between the time of original forest removal and Spanish Contact may contribute to these differences (Leyden 1987; Jones 1991). To control these possibilities, Piperno et al. (1991b) compared the species composition of the original preremoval forest at La Yeguada, Panama, the post-Contact recovering forest at the same site, and modern forest in central Panama that was never cleared for agriculture. Several tree taxa were found only in the original preremoval forest at La Yeguada and in the undisturbed modern forest. As a vegetation type, forest is resilient; however, the species composition of tropical forests is not particularly resilient.

Finally, the stability of tropical forests may be moot. Given the long life spans of the dominant plant life forms, most modern tropical forests may be recovering from recent, massive disturbances. Demographic extrapolations from 8-year records of growth and survivorship of 300 000 tropical trees suggest that the life spans of individuals of mature forest species range from 200 to 800 years (R. Condit, pers. comm.). For a species with a generation time at the lower end of this range (200 years), rapid glacial climate change occurred just 50 generations ago, massive forest removal by indigenous peoples ended just 2.5 generations ago over large parts of the Neotropics, and fire associated with severe droughts may occur once every two generations (Woods 1989). Our impression of tropical forest stability may have more to do with our own short life spans than with constancy in the species composition and abundances of forest plants.

2.6 Conclusions

The relationship between species richness and ecosystem functioning depends on interspecific interactions. If each species performs well under a unique set of environmental conditions with little overlap among species and if environmental conditions also vary in space or time, then ecosystem productivity and stability should increase with species richness (McNaughton 1993). If, on the other hand, there is broad overlap among species and a high degree of functional redundancy, then ecosystem productivity and stability should be independent of species richness. This latter possibility may describe tropical forest plant communities.

Studies of the performance of tropical forest plants along a variety of environmental gradients indicate that many species have broad environmental tolerances and broadly overlapping niches. Examples include plant responses to gradients in light associated with treefall gaps (Denslow et al. 1990; Welden et al. 1991), gradients in soil drainage (Lieberman et al. 1985; Hubbell and Foster 1986b), and gradients in soil fertility (Baillie et al. 1987). In each example, large numbers of species were either indifferent to or, in the case of light, shared similar responses to the environmental gradient. Such species have broad ecological overlap. A high level of functional redundancy follows at the high species richness typical of tropical forests.

For these reasons, ecosystem processes should reach an asymptote with respect to plant species richness at levels of plant species richness well below those commonly observed in tropical forests. This prediction was upheld in three comparisons involving forest productivity. Fine litter production was similar in highly dominated and highly diverse tropical forests (Fig. 2.3), wood production was similar in tropical plantations and native forests (Jordan 1983), and, in the one relevant experimental study, forest leaf area and root biomass were unaffected by plant species density (Brown and Ewel 1987; Berish and Ewel 1988). These studies indicate that tropical forest productivity is largely independent of species richness, at least over the ecologically relevant range of species richnesses observed in tropical forests.

The apparent stability of species-rich tropical forests was critical to the early development of theory that predicted that ecosystem stability should increase with species diversity (Elton 1958). Subsequent studies have shown that tropical forests are, in fact, not particularly stable with respect to population variability and perhaps invasibility. Although there have been no controlled studies of tropical forest stability, a series of observations and natural experiments reviewed here provided no evidence that tropical forests are uniquely resistant or resilient to a variety of perturbations including hurricanes, fire, climate change, and anthropogenic removal.

Tilman and Downing (1994) recently demonstrated that greater species richness increased the resistance to drought and resilience of prairie plant communities. Although their study clearly establishes a link between plant species richness and the stability and resilience of ecosystem productivity, the details again suggest that this link may have limited relevance to tropical forests. Tilman and Downing (1994) observed increased productivity and more rapid recovery over the range from one to just ten species. An asymptote was reached for communities with more than ten species. In contrast, a drought in central Panama that might be expected once each century also revealed the species specific responses that might lead to an increase in ecosystem stability with species richness. However, in Panama, although several dozen tree species suffered increased mortality, more than 100 species did not (Hubbell and Foster 1990). Plant species richness clearly affects ecosystem productivity in species-poor communities, however, this relationship is unlikely to be important at the high species richness typical of tropical forests.

In a recent review of biogeochemical processes in tropical forests, Vitousek and Hooper (1993) concluded that plant diversity is most likely to affect ecosystem functioning over the range from one to ten species. A similar conclusion may well be true for productivity and stability also, although the critical range of species richness may be somewhat higher. This conclusion should obviously be regarded as a hypothesis at this time. The long life spans of the dominant plant life forms make it unlikely that the hypothesis will be tested directly with controlled experiments in mature tropical forests. Nevertheless, tests are possible. McNaughton (1985, 1993) has performed a powerful series of tests of the relation between Serengeti plant diversity and ecosystem stability using a combination of short-term experimental manipulations and comparative analyses. Tropical forests where one or a few tree species are dominant are often contiguous with forests where tree diversity is high. In some instances, these boundaries do not coincide with edaphic or climatic differences (Hart *et al.* 1989; Hart 1990). An approach similar to McNaughton's applied across such a boundary might yet identify ecosystem dependence on plant diversity that is not apparent in the coarser comparative analyses possible at this time.

References

Addicott FT (1982) Abscission. Univ California Press, Berkeley, California

Allee WC (1926) Measurement of environmental factors in the tropical rain-forest of Panama. Ecology 7:273-302

Andrews RM, Wright SJ (1994) Long-term population fluctuations of a tropical lizard: a test of causality. In: Vitt L, Pianka ER (eds) Lizard ecology: historical and experimental perspectives. Princeton Univ Press, Princeton, New Jersey, pp 267-285

Ashton PS (1969) Speciation among tropical forest trees: some deductions in the light of recent evidence. In: Lowe-McConnell RH (ed) Speciation in tropical environments. Academic Press, London, pp 155-196

Ashton PS (1977) A contribution of rain forest research to evolutionary theory. Ann Mo Bot Gard 64:694-705

Ashton PS (1988) Dipterocarp biology as a window to the understanding of tropical forest structure. Annu Rev Ecol Syst 19:347-370

Ashton PS (1993) Species richness in plant communities. In: Fiedler PL, Jain SK (eds) Conservation biology. Chapman and Hall, New York, pp 4-22

Austin MP, Smith TM (1989) A new model for the continuum concept. Vegetatio 83:35-47

Baillie IC, Ashton PS, Court MN, Anderson JAR, Fitzpatrick EA, Tinsley J (1987) Site characteristics and the distribution of tree species in mixed dipterocarp forest on Tertiary sediments in central Sarawak, Malaysia. J Trop Ecol 3:201-220

Berish CW, Ewel JJ (1988) Root development in simple and complex tropical successional ecosystems. Plant Soil 106:73-84

Bjorkman O, Demmig B (1987) Photon yield of O_2 evolution and chlorophyll fluorescence characteristics at 77 K among vascular plants of diverse origins. Planta 170:489-504

Blackwell WH Jr (1968) Flora of Panama, Part VIII. Family 154. Sapotaceae. Ann Mo Bot Garden 55:145-169

Bock IR (1980) Current status of the *Drosophila melanogaster* species group. Syst Entomol 5:341-356

Brokaw N (1987) Gap-phase regeneration of three pioneer tree species in a tropical forest. J Ecol 75:9-19

Brokaw N, Walker LR (1991) Summary of the effects of Caribbean hurricanes on vegetation. Biotropica 23:442-447

Brown BJ, Ewel JJ (1987) Herbivory in complex and simple tropical successional ecosystems. Ecology 68:108- 116

Brown S, Lugo AE (1982) The storage and production of organic matter in tropcal forests and their role in the global carbon cycle. Biotropica 14:161-187

Brown S, Lugo AE (1990) Tropical secondary forests. J Trop Ecol 6:1-32

Bush MB, Colinvaux PA (1990) A pollen record of a complete glacial cycle from lowland Panama. J Veg Sci 1:105-118

Bush MB, Colinvaux PA (1994) Tropical forest disturbance: paleoecological records from Darien, Panama. Ecology 75:1761-1768

Bush MB, Piperno DR, Colinvaux PA, De Oliveira PE, Krissek LA, Miller MC, Rowe WE (1992) A 14,300-yr paleoecological profile of a lowland tropical lake in Panama. Ecol Monogr 62:251-275

Cavelier J, Wright SJ (1995) Effects of irrigation on fine root biomass and production, litterfall and trunk growth in a semideciduous lowland forest in Panama. J Trop Ecol (in press)

Connell JH, Lowman MD (1989) Low-diversity tropical rain forests: some possible mechanisms for their coexistence. Am Nat 134: 88-119

Cuevas E, Medina E (1988) Nutrient dynamics within Amazonian forests. II. fine root growth, nutrient availability and leaf litter decomposition. Oecologia 76:222-235

Denslow JS (1980) Gap partitioning among tropical rain forest trees. Biotropica 12(Suppl):47-55

Denslow JS (1987) Tropical rainforest gaps and tree species diversity. Annu Rev Ecol Syst 18:431-452

Denslow JS, Schultz JC, Vitousek PM, Strain BR (1990) Growth responses of tropical shrubs to treefall gap environments. Ecology 71:165-179

Elton CS (1958) The ecology of invasions by animals and plants. Methuen, London

Foster RB (1982) Famine on Barro Colorado Island. In: Leigh EG Jr, Rand AS, Windsor DM (eds) Ecology of a tropical forest. Smithsonian Inst Press, Washington DC, pp 201-211

Garwood NC (1989) Tropical soil seed banks: a review. In: Leck MA, Parker VT, Simpson RL (eds) Ecology of soil seed banks. Academic Press, New York, pp 149-209

Gentry AH (1986) Endemism in tropical versus temperate plant communities. In: Soulé ME (ed) Conservation biology. Sinauer, Sunderland, MA, pp 153-181

Gentry AH (1988a) Tree species richness of upper Amazonian forests. Proc Natl Acad Sci USA 85:156-159

Gentry AH (1988b) Changes in plant community diversity and floristic composition on environmental and geographical gradients. An Mo Bot Gard 75:1-34

Gentry AH, Dodson C (1987) Contribution of nontrees to species richness of a tropical rain forest. Biotropica 19:149-156

Gentry AH, Emmons LH (1987) Geographical variation in fertility, phenology, and composition of the understory of Neotropical forests. Biotropica 19:216-227

Giacalone J, Glanz WE, Leigh EG Jr (1990) Fluctuaciones poblaciones a largo plazo de *Sciurus granatensis*. In: Leigh EG Jr, Rand AS, Windsor DM (eds) Ecología de un bosque tropical. Smithsonian Inst Press, Washington DC, pp 455-468

Gilbert GS, Hubbell SP, Foster RB (1994) Density and distance-to-adult effects of a canker disease of trees in a moist tropical forest. Oecologia 98:100-108

Glanz WE (1982) The terrestrial mammal fauna of Barro Colorado Island. In: Leigh EG Jr, Rand AS, Windsor DM (eds) Ecology of a tropical forest. Smithsonian Inst Press, Washington DC, pp 455-468

Golley FB (1969) Caloric value of wet tropical forest vegetation. Ecology 50:517-519

Gresham CA, Williams TM, Lipscomb DJ (1991) Hurricane Hugo wind damage to southeastern US coastal forest tree species. Biotropica 23:420-426

Hall JB, Swaine MD (1981) Distribution and ecology of vascular plants in a tropical rain forest: forest vegetation in Ghana. Junk, The Hague, Geobotany 1:383

Hamilton AC (1982) Environmental history of East Africa. Academic Press, New York, 328 pp

Hart TB (1990) Monospecific dominance in tropical rain forests. Trends Ecol Evol 5:6-11

Hart TB, Hart JA, Murphy PG (1989) Monodominant and species-rich forests of the humid tropics: causes for their co-occurrence. Am Nat 133:613-633

Hart TB, Hart JA, Dechamps R, Fournier M, Ataholo M (1995) Changes in forest composition over the last 4000 years in the Ituri Basin, Zaire. In: Proc 14th Congr of the Association for the Taxonomic Study of the Flora of Tropical Africa (in press)

Holdridge LR, Grenke WC, Hatheway WH, Liang T, Tosi JA Jr (1971) Forest environments in tropical life zones. Pergamon Press, Oxford

Hubbell SP, Foster RB (1986a) Biology, chance, and history, and the structure of tropical rain forest. In: Diamond J, Case TJ (eds) Community ecology. Harper & Row, New York, pp 314-329

Hubbell SP, Foster RB (1986b) Commonness and rarity in a neotropical forest: implications for tropical tree conservation. In: Soulé M (ed) Conservation biology: science of scarcity and diversity. Sinauer, Sunderland, MA, pp 205-231

Hubbell SP, Foster RB (1990) Structure, dynamics and equilibrium status of old-growth forest on Barro Colorado Island. In: Gentry A (ed) Four neotropical forests. Yale Univ Press, New Haven, Connecticut, pp 522-541

Huston MA (1980) Soil nutrients and tree species richness in Costa Rican forests. J Biogeogr 7:147-157

Huston MA (1994) Biodiversity. Cambridge Univ Press, Cambridge

Janzen DH (1972) *Jacquinia pungens*, a heliophile from the understory of tropical deciduous forest. Biotropica 2:112-119

Janzen DH (1981) Patterns of herbivory in a tropical deciduous forest. Biotropica 13:271-282

Janzen DH (1984) Natural history of *Hylesia lineata* (Saturniidae: Hemileucinae) in Santa Rosa National Park, Costa Rica. J Kansas Entomol Soc 57:490-514

Janzen DH (1985) A host plant is more than its chemistry. Ill Nat Hist Surv Bull 33:141-174

Jones JG (1991) Pollen evidence of prehistoric forest modification and Maya cultivation in Belize. PhD Dissertation, Texas A&M Univ, Lubbock, Texas, 131 pp

Jordan CF (1983) Productivity of tropical rain forest ecosystems and the implications for their use as future wood and energy sources. In: Golley FB (ed) Ecosystems of the world 14A tropical rain forest ecosystems. Elsevier, Amsterdam, pp 117-136

Jordan CF, Escalante G (1980) Root productivity in an Amazonian rain forest. Ecology 61:14-18

Kira T (1978) Community architecture and organic matter dynamics in tropical lowland rain forests of Southeast Asia with special reference to Pasoh forest, West Malaysia. In: Tomlinson PB, Zimmermann MH (eds) Tropical trees as living systems. Cambridge Univ Press, Cambridge, pp 561-590

Kira T, Ogawa H (1971) Assessment of primary production in tropical and equatorial forests. In: Duvigneaud P (ed) Productivity of forest ecosystems. UNESCO, Paris, pp 309-321

Leyden BW (1987) Man and climate in the Maya lowlands. Quat Res 28:407-414

Lieberman M, Lieberman D, Hartshorn GS, Peralta R (1985) Small-scale altitudinal variation in lowland wet tropical forest vegetation. J Ecol 73:505-516

Lugo AE (1992) Comparison of tropical tree plantations with secondary forests of similar age. Ecol Monog 62:1-41

MacArthur RH, Wilson EO (1967) The theory of island biogeography. Princeton Univ Press, Princeton, New Jersey

May RM, MacArthur RH (1972) Niche overlap as a function of environmental variability. Proc Natl Acad Sci USA 69:1109-1113

McNaughton SJ (1985) Ecology of a grazing ecosystem: the Serengeti. Ecol Monogr 55:259-294

McNaughton SJ (1993) Biodiversity and function of grazing ecosystems. In: Schulze E-D, Mooney HA (eds) Biodiversity and ecosystem function. Springer, Berlin Heidelberg New York, pp 361-382

Medina E, Klinge H (1983) Productivity of tropical forests and tropical woodlands. In: Lange OL, Nobel PS, Osmond CB, Ziegler H (eds) Physiological plant ecology IV. Springer, Berlin Heidelberg New York, pp 281-303

Muller D, Nielsen J (1965) Production brute, pertes par respiration et production nette dans la foret ombrophile tropicale. Forstl Forsogsvaes Dan 29:60-160

Naeem S, Thompson LJ, Lawler SP, Lawton JH, Woodfin RM (1994) Declining biodiversity can alter the performance of ecosystems. Nature 368:734-737

Odum HT (ed) (1970) A tropical rain forest. Div Tech Info US Atomic Energy Comm, Washington, DC

Ogawa H, Yoda K, Ogino K, Kira T (1965) Comparative ecological studies on three main types of forest vegetation in Thailand II. plant biomass. Nat Life Southeast Asia 4:49-80

Paine RT (1966) Food web complexity and species diversity. Am Nat 100:65-75

Pimm SL (1984) The complexity and stability of ecosystems. Nature 307:321-326

Piperno DR, Becker P (1995) Vegetation changes and fire in central Amazonian rain forest during Holocene dry periods. Ecology (in review)

Piperno DR, Bush MB, Colinvaux PA (1991a) Paleoecological perspectives on human adaptaton in central Panama I. The Pleistocene. Geoarchaeology 6:201-226

Piperno DR, Bush MB, Colinvaux PA (1991b) Paleoecological perspectives on human adaptaton in central Panama II. The Holocene. Geoarchaeology 6:227-250

Prentice IC, Bartlein PJ, Webb T III. (1991) Vegetation and climate change in eastern North America since the last glacial maximum. Ecology 72:2038-2056

Proctor J (1984) Tropical forest litterfall II: the data set. In: Chadwick AC, Sutton SL (eds) Tropical rain-forest: the Leeds symposium. Leeds Philosophical and Literary Society, Leeds, UK, pp 83-113

Reich PB, Borchert R (1984) Water stress and tree phenology in a tropical dry forest in the lowlands of Costa Rica. J Ecol 72:61-74

Richards PW (1952) The tropical rain forest. Cambridge Univ Press, Cambridge, 450 pp

Ricklefs RE (1977) Environmental heterogeneity and plant species diversity: a hypothesis. Am Nat 111:376-381

Roubik DW (1991) Aspects of Africanized honey bee ecology in tropical America. In: Spivak M, Fletcher DJC, Breed MD (eds) The "African" honey bee. Westview Press, Boulder, Colorado, pp 259-281

Roubik DW (1993) Direct costs of forest reproduction, bee-cycling and the efficiency of pollination modes. J Biosci 18:537-552

Sanford RL Jr, Saldarriaga J, Clark KE, Uhl C, Herrera R (1985) Amazon rain-forest fires. Science 227:53-55

Sevenstar JG (1991) The community ecology of frugivorous Drosophila in a Neotropical forest. PhD Dissertatio, Rijksuniv te Leiden, Holland

Smith AP (1987) Respuestas de hierbas del sotobosque tropical a claros ocasionados por la caida de arboles. Rev Biol Trop 35 (Suppl 1):111-118

Terborgh J (1986) Keystone plant resources in the tropical forest. In: Soulé ME (ed) Conservation biology. Sinauer, Sunderland, MA, pp 330-344

Tilman D, Downing JA (1994) Biodiversity and stability in grasslands. Nature 367:363-365

Valencia R, Balslev H, Paz y Mino C G (1994) High tree alpha-diversity in Amazonian Ecuador. Biodiv Conserv 3:21-28

Van Schaik CP, Terborgh JW, Wright SJ (1993) The phenology of tropical forests: adaptive significance and consequences for primary consumers. Annu Rev Ecol Syst 24:353-377

Vitousek PM, Hooper DU (1993) Biological diversity and terrestrial ecosystem biogeochemistry. In: Schulze E-D, Mooney HA (eds) Biodiversity and ecosystem function. Springer, Berlin Heidelberg New York, pp 3-14

Welden CW, Hewett SW, Hubbell SP, Foster RB (1991) Sapling survival, growth, and recruitment: relationship to canopy height in a neotropical forest. Ecology 72:35-50

Wolda H (1978) Seasonal fluctuations in rainfall, food, and abundance of tropical insects. J Anim Ecol 47:369-381

Wolda H (1988) Insect seasonality: why? Annu Rev Ecol Syst 19:1-18

Wolda H, Foster RB (1978) Zunacetha annulata (Lepidoptera: Dioptidae), an outbreak insect in a Neotropical forest. Geo-Eco-Trop 2:443-454

Wong M, Wright SJ, Hubbell SP, Foster RB (1990) The spatial pattern and reproductive consequences of outbreak defoliation in Quararibea asterolepis, a tropical tree. J Ecol 78:579-588

Woods P (1989) Effects of logging, drought, and fire on structure and compositon of tropical forests in Sabah, Malaysia. Biotropica 21:290-298

Wright SJ (1992) Seasonal drought, soil fertility and the species density of tropical forest plant communities. Trends Ecol Evol 7:260-263

Wright SJ, Cornejo FH (1990) Seasonal drought and leaf fall in a tropical forest. Ecology 71:1165-1175

Wright SJ, Van Schaik CP (1994) Light and the phenology of tropical forest trees. Am Nat 143:192-199

3 Consumer Diversity and Secondary Production[1]

Michael Huston[2] and Larry Gilbert[3]

3.1 Introduction

Thirty years of ecosystem research have demonstrated unequivocally that certain species of plants, microbes, or animals can have dramatic effects on such ecosystem processes and properties as primary productivity, soil chemistry and structure, evapotranspiration, leaf area and fluxes of trace gases. However, the significance of having a particular number of species participating in a particular ecosystem process or continuum of ecosystem processes is less obvious. Apparently well-functioning natural ecosystems can be found in which most of the biomass is composed of one or two species of plants or other functional groups. Likewise, many ecosystems apparently maintain full functionality despite major changes in the identities and total number of species over time. Thus it is difficult to conclude that biodiversity *per se* has any particular significance for ecosystem processes. Many species are not necessarily better than few species for carrying out or maintaining a particular ecosystem process.

Extensive evidence demonstrates that environmental conditions regulate biodiversity. Biodiversity varies predictably along elevational, disturbance, productivity, moisture, and latitudinal gradients (Huston, 1994).

Although the diversity of many different taxa reaches a maximum in lowland tropical rain forests, the reasons for these diversity maxima vary significantly among taxa. In fact, the response to any environmental factors known to influence diversity is likely to vary among taxa, with some taxa increasing in diversity and others decreasing in diversity in response to a single gradient or change in conditions. Biodiversity patterns can best be understood if they are sub-

[1] The manuscript has been authored by a contractor of the U.S. Government under contract No. DE-AC05-84OR21400. Accordingly, the U.S. Government retains a nonexclusive, royalty-free licence to publish or reproduce the published form of this contribution, or allow others to do so, for U.S. Government purposes

[2] Environmental Division, Oak Ridge National Laboratory, P. O. Box 2008, Oak Ridge, Tennessee 37831-6038 USA

[3] Department of Zoology, University of Texas, Austin, Texas 78712, USA

Ecological Studies, Vol. 122
Orians, Dirzo and Cushman (eds) Biodiversity and Ecosystem Processes in Tropical Forests
© Springer-Verlag Berlin Heidelberg 1996

divided into groups of species that are reasonably consistent in their inter-actions with each other, with other species, and with the physical environment. Such groups, which have been described variously as guilds, strategies, growth forms, and functional types, have long been an important concept in ecology. A major rationale for this subdivision of organisms is that interactions such as competition are more likely to be stronger among organisms of the same guild or functional type than among organisms of different functional types.

3.2 Secondary Production and Biodiversity

Secondary production is the amount of biomass per unit area per unit time that is produced by organisms that consume, directly or indirectly, the biomass produced by plants. Plants influence animal biodiversity through two primary mechanisms. First, plants provide the energy that supports animal populations and allows them to maintain enough individuals and have a sufficient rate of growth and recovery that the populations can persist through natural fluctua-tions and severe disturbances. Second, plants provide physical, temporal, and biochemical heterogeneity that favors the evolution of animals that specialize on different plant parts and plant species, thereby reducing interspecific com-petition and allowing many animal species to coexist in the same area.

These two mechanisms seem to predict a world where the highest diversity of all organisms would be found in those environments where both primary and secondary productivity reached their maximum. However, contrary to these predictions, plant diversity tends to be low under environmental conditions where plant productivity is very high and the maximum plant diversity gener-ally occurs under poor environmental conditions where plant productivity is quite low. The apparent cause of reduced plant diversity under the most pro-ductive conditions is competitive exclusion resulting from competition for light, in which a few large or rapidly growing species shade and eliminate smaller or more slowly growing species. This same phenomenon also seems to be the ex-planation for the reduction in diversity that occurs in late succession in many different ecosystems (Huston 1994). This pattern is found among competing plant species in experimental nutrient manipulations and along natural pro-ductivity gradients in virtually all ecosystems of the world (Fig. 3.1).

Thus, the two primary mechanisms by which plants influence animal diver-sity often may be inversely correlated, with plant diversity being high in low productivity environments, and plant productivity high in environments where plant diversity tends to be low. The implications of this pattern for species rich-ness of animals are complex. Those groups of animals that tend to specialize on different plant species or plant parts, including most insects and some birds and bats, would be expected to reach their highest diversity where plant diversity is highest (Fig. 3.2A).

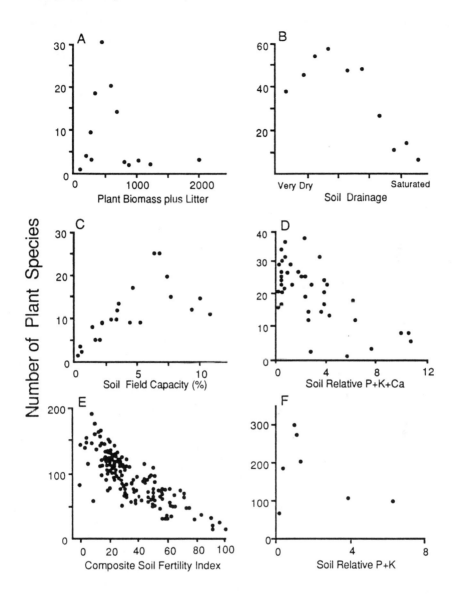

Fig. 3.1. Association of plant species number with primary productivity. *A* British herbaceous plants; productivity indexed by sum of living and dead plant biomass. *B* North American prairie plants; productivity indexed by soil drainage. *C* Annual plants in an Israel desert; plant productivity indexed by soil water-holding potential. *D* Number of Costa Rican rain forest trees per 0.1 ha; soil fertility indexed as the sum of relative soil P, K, and Ca among 46 sites. *E* Number of vascular plants per 0.1 ha in Ghanian forests; soil fertility indexed by a principal components axis negatively correlated with total exchangeable bases and other measures of soil fertility. *F* Number of trees per ha in West Malesian rain forest; soil fertility indexed as the sum of relative soil P and K. (After Huston 1994)

In contrast, those groups of animals that are more generalized in their plant use and depend on plants primarily as an energy source, such as many vertebrate herbivores and the higher trophic levels that they support, as well as the detrital and decomposer communities, would be expected to reach their highest diversity where plant productivity is highest, independent of plant diversity (Fig. 3.2 B). These expected patterns of trophic level diversity are summarized in Fig. 3.3, which is based on the fact that energetic losses between trophic levels mean that population sizes and growth rates will generally be smaller at high trophic levels than at low trophic levels, causing a shift in the maximum diversity of high trophic levels to higher primary productivity.

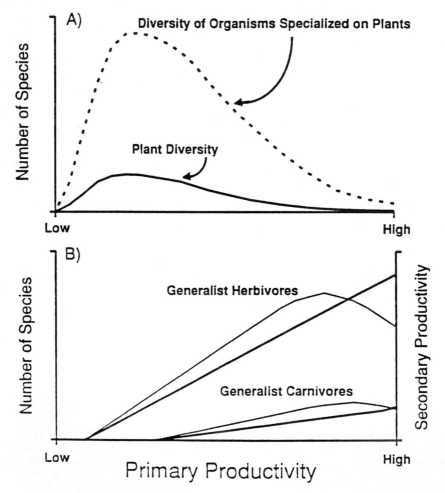

Fig. 3.2. Expected patterns of diversity among specialist versus generalist organisms. **A** Diversity of organisms having host-specialist interactions with plants. **B** Diversitiy of organisms that depend on total plant productivity. (after Huston 1994)

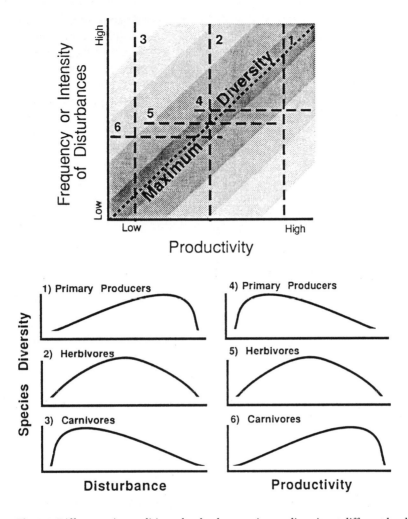

Fig. 3.3. Differences in conditions that lead to maximum diversity at different levels in a food web as a result of energy loss with each transfer between levels in the food chain. (After Huston 1994)

However, secondary production is not a simple function of primary production because plants have evolved a number of defensive strategies, including toxic chemicals, that deter herbivores, parasites, and pathogens from consuming plant material, and also prevent those organisms from efficiently utilizing the plant material that they do consume. Thus, not all of the energy and other resources contained in primary productivity are available to consumers. Secondary production and its relationship to primary productivity are extremely difficult to measure. Many consumers, particularly microbes and insects, are extremely small, others are mobile and difficult to sample systematically. As a result, there are few good data on secondary production, particularly total secondary production of entire ecosystems. However, data on species richness at higher trophic levels in arid ecosystems show exactly the pattern predicted by the "trophic shift" hypothesis (Huston 1992, 1994), although at a very large scale. In arid ecosystems, water availability is the main limiting factor for primary production, so that precipitation gradients tend to also be productivity gradients. In Texas, Owen (1988) found that rodent diversity (number of species recorded in 189 grid cells, each with an area of 4082 km^2) reached its maximum at an estimated primary productivity of about 150 (g m^{-2}yr^{-1}) whereas carnivore diversity peaked at an estimated primary productivity of 1400 (g m^{-2} yr^{-1}). Perhaps coincidentally, this order of magnitude difference in primary productivity at which the diversity of rodents and their carnivores reaches a maximum is equivalent to the energy lost in transfer between these two trophic levels, a value that has been documented in a number of trophic dynamic studies (e.g., Golley 1960). This suggests that the diversity of these two general functional types of vertebrates may actually reach maximum diversity at the same level of trophically adjusted secondary productivity.

Few comparative data are available on the relation between primary and secondary production in tropical rain forests, or between primary and secondary production and animal diversity. Data from four intensively studied neotropical rain forests (Gentry 1990) are difficult to compare quantitatively because of different sampling methods and sample sizes. However, the highest numbers of nonflying mammal species and total carnivore species (birds and mammals) among the four sites are found at the Cocha Cashu reserve on the fertile alluvial floodplain of the Manu River.

The decomposer community of microinvertebrates reaches extremely high diversity under conditions of high plant productivity and organic matter accumulation in the coniferous forests of the Pacific Northwest of North America (Lattin 1990), where the diversity of tree species is very low. In contrast to the microinvertebrate decomposers, among which energy supply seems more important than host specificity, recent evidence suggests that host specificity may be even higher among decomposing fungi than among fungi of living plants (J. Lodge, pers. comm.). This suggests that fungal diversity may be extremely high in tropical rain forests with high plant diversity. Unfortunately, few data are available for quantitative comparisons of decomposer diversity along gradients of primary productivity and substrate diversity, particularly in tropical forests.

3.3 Evolutionary Effects of Consumers on Ecosystem Properties

Available primary production is clearly a critical determinant of the secondary productivity, and perhaps diversity, of higher trophic levels. Somewhat surprisingly, available data suggest that annual primary productivity is not much higher in tropical forests than in many temperate forests (Jordan 1983; Huston 1994). A significant difference between humid lowland tropical forests and temperate forests is that production occurs throughout the year in the tropics, but is concentrated in a relatively short growing season in temperate forests.

The mild, predictable growing conditions of relatively aseasonal rain forests allow more constant herbivore pressure, which leads to increased allocation of plant resources to chemical defenses (e.g., condensed tannins) in tropical foliage (Coley and Aide 1988) and probably to increased diversification of low molecular weight toxins that can be mobilized to protect new growth (McKey 1979). Many plants, such as *Passiflora* vines, have potent chemical defenses that exclude all but a few groups of dedicated metabolic specialist insects from eating their foliage. When chemical defenses prove ineffective, other factors, such as harsh abiotic conditions or predators and parasites, may effectively prevent specialist consumers, such as the *Passiflora* feeders, from severely damaging their host plants. Consequently, plants are also under evolutionary pressure to supplement their chemical and physical defenses with inducements to attract predatory and parasitic insects, such as ants and wasps. Extrafloral nectaries and food bodies (which attract ants, wasps, and parasitoids) increase dramatically along a gradient from temperate to tropical regions (Coley and Aide 1988).

These diffuse, multispecies coevolutionary interactions result in plants allocating a significant amount of energy and resources to two higher trophic levels: (1) chemical defenses to deter herbivores, and (2) energy and other resources used to attract predators and parasites of the herbivores. This allocation to defense reduces both primary and secondary production (Simms 1992). Plants selected by agriculturists to reduce their allocation to chemical defenses are able to outproduce their wild relatives under conditions where herbivores and pathogens are controlled by human-made pesticides. Thus, we may conclude that these coevolutionary plant-herbivore interactions have reduced the secondary production to some degree.

Furthermore, the assimilation efficiency of the consumer trophic level is inevitably reduced by the costs of detoxifying or otherwise countering plant defenses. The evolution of variety in plant chemical defenses under pressure from bacteria, fungi, nematodes, insects, and herbivorous vertebrates has promoted evolutionary diversification of these consumers. Multi-species coevolution has created many highly specific pathways by which primary production is converted to secondary production. Some of these pathways transfer a much higher proportion of total plant energy to higher trophic levels than do other pathways.

For example, in spite of some reduction in assimilation efficiency due to the cost of neutralizing plant chemical defenses, herbivorous insects have an assimilation efficiency of ~40% (Weigert and Evans 1967). In contrast, the need for and cost of thermoregulation lowers the net assimilation efficiency of herbivorous mammals to the 1.5-3.0% range. This limitation forces mammalian herbivores to be larger and more generalized in feeding habits than insects, and to depend heavily on microbial detoxification systems. Consequently, both the diversity and productivity of invertebrate herbivores should be more strongly correlated with the diversity and productivity of plants than are the same attributes of vertebrate herbivore systems.

Within tropical forests there is some evidence that allocation to plant defenses is higher in forests where primary production is lower (McKey et al. 1978). Thus the already low amount of energy available to support higher trophic levels in these unproductive systems is further reduced by the energetic costs of evolved antiherbivore defenses. These patterns within the tropics are consistent with the apparent global-scale latitudinal gradient of decreasing productivity in the tropics (Huston 1993) that is inversely related to the gradient of increasing allocation to plant defenses in the tropics (Coley and Aide 1988).

The generally low primary productivity of most lowland humid tropical forests also makes resource concentrations particularly important. It is under these conditions that certain plant species can serve as "keystone resources," providing a critical food source that allows many species of animals to survive periods of low food availability (Terborgh 1986). The productivity/disturbance conditions that influence species diversity also define the conditions under which different types of "keystone species" are likely to occur and evolve (Fig. 3.4).

3.4 Ecological Effects of Consumers on Ecosystem Properties

The above arguments describe patterns of consumer productivity and biodiversity as the inevitable ecological and evolutionary consequences of the physical environmental conditions that produce a particular combination of primary productivity and plant diversity. Nonetheless, it is clear that consumers can have a major short-term impact on ecosystem processes as well as a long-term evolutionary effect as partners in coevolutionary interactions and selective forces driving plant defensive and reproductive adaptations.

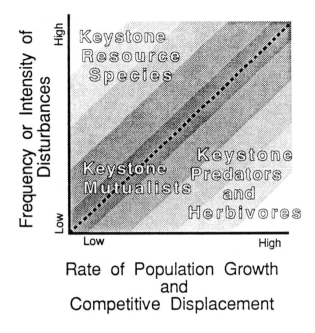

Fig. 3.4. Conditions of productivity and disturbance that favor different types of keystone species. (After Huston 1994)

3.4.1 Influence of consumers on plant productivity

Consumers can potentially act to increase (or decrease) plant biodiversity and plant productivity. In a number of different systems, notably planktonic or sessile algae communities and terrestrial grasslands, consumers can increase primary production by maintaining individual plants or plant populations in a phase of rapid growth. The accumulation of living plant biomass, which occurs over time in the absence of harvesting, decreases net primary production as a result of respiration and by tying up essential nutrients in the biomass. Consumers can potentially increase NPP by reducing respiratory losses and increasing the release and recycling of nutrients, both of which result from consumption of plant biomass. In forested ecosystems, where the amount of standing biomass is high in relation to leaf area, there is no evidence that the relatively small amount of NPP (5-10%) consumed by herbivores has any effect on total NPP (Coley and Aide, 1988). Thus in tropical rain forest systems it seems unlikely that herbivores have much direct negative effect on total primary production.

One reason that herbivores have little direct effect on primary production may be the control exerted by inconspicuous consumers such as blood parasites of vertebrates, predatory mites, and microhymenoptera (parasitoid wasps). These tiny organisms can act as rate regulators or "energy filters" (Hubbell 1973) by controlling the population of herbivores and thus reducing the rate and number of pathways by which primary production becomes secondary production. Microhymenoptera are thought to be a principal factor reducing insect herbivory in polyculture crop systems (Andow 1983) as well as natural systems (e.g., Gilbert 1977). Likewise, the low density, specialist insects of a plant are unlikely to be identified as keystone predators until the plant becomes an introduced weed in another ecosystem and one or more of its specialist fauna are imported to serve as "magic bullets" in its biological control. Such components of secondary productivity and overall biodiversity are basically ignored by the functional ecosystem approach, yet the system cannot be understood without knowing more details about them. The study of introduced species and agricultural systems should provide insights into the influence of biodiversity on ecosystem functioning (Vitousek 1986, Lawton and Brown 1993). The most important components of that biodiversity may be typically rare or inconspicuous organisms.

Potential interactions between all trophic levels result in a complex system of controls and feedbacks in the fluxes and cycling of energy and nutrients through ecosystems. The complexity and biological richness of these interactions is suggested in Fig. 3.5 which highlights a few of the better-understood interactions from neotropical rain forests and illustrates the direct and indirect effects that all higher trophic levels have on primary production. Herbivorous insect larvae (e.g., caterpillars), which are important components of consumer categories (Fig. 3.5, 14, 15), are responsible for a substantial fraction of energy transfer from primary producers to secondary consumers (Fig. 3.5A). Microhymenoptera (Fig. 3.5, 20), in spite of being inconspicuous and possessing negligible biomass, can dramatically affect (or "filter") the flow of leaf energy into herbivores by killing insect herbivore eggs.

Herbivorous insect larvae and adults are attacked by both specialist and generalist invertebrate predators (Fig. 3.5, 20, 21). The principal impact of this mortality is in density-dependent control of prey that maintains herbivorous insect populations below outbreak levels (Fig. 3.5B). Social wasps and ants (Fig. 3.5, 21), which are important mortality agents for foliage-feeding insects, are themselves victimized by army ants (Fig. 3.5, 27) on a regular basis. The outcomes of these interactions may be a reduction of variance in the flow of energy from plants to the herbivore trophic levels and a promotion of genetic diversity (in terms of predator escape strategies) in the consumer community.

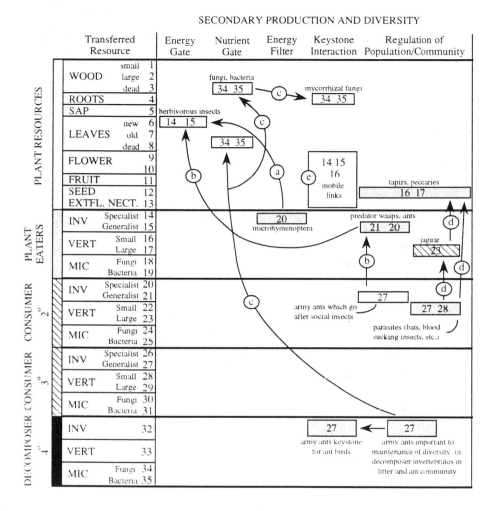

Fig. 3.5. Interactions that affect ecosystem processess and/or biodiversity in tropical rain forests. Categories of resources consumed and consumers are organized according to trophic level, major taxa, and life history groups (*left column*). These 35 functional groups (the *rows*) divide the ecosystem into biotic units that make sense in population, community, and evolutionary ecology, and provide a framework for relating ecosystem and population biology. Sample interactions involving consumers that affect ecosystem processes such as energy and nutrient flow are indicated in the resource x ecosystem function cells and by *arrows* from other parts of the chart to those cells. *Numbers in the cells* indicate example groups of consumers that may be involved. *Arrows* come from cells in which high order consumers influence the activities of lower order consumers. Interactions principally affecting population and community properties are indicated under the *Keystone* and *Regulation* columns. If there is a relation between biodiversity and ecosystem functioning in tropical forests, it should be found in interactions networks (*cells connected by arrows*) such as those illustrated. The specific neotropical examples (a - e) are discussed in the text.

The nutrient cycle depends on the activities of fungi and bacteria (Fig. 3.5, 34, 35); (Fig. 3.5C). Dead plant matter is heterogeneous, and a variety of invertebrate decomposers (Fig. 3.5, 32) is required for maximum nutrient turnover. Army ants (Fig 3.5, 27) may play a role in maintaining the diversity of litter arthropods through the disturbance caused by their predatory activities in the litter community. Army ants also play a keystone role for the ant-bird community.

3.4.2 Influences of Consumers on Plant Diversity

The potential diversifying effects of predators on their prey, and herbivores on plants are well known. Generalized predators, such as lawnmowers (Darwin 1859), can have a diversifying effect by removing a larger proportion of biomass from the most abundant species or by affecting all species equally (Huston 1979). Specialized predators can have a diversifying effect if they reduce populations of the most abundant species more than rare species (Connell 1971, Janzen 1970). Thus, under some conditions, consumers can increase the diversity of plants, which should lead to higher diversity among those consumers dependent on plant diversity.

Vertebrate seed and seedling predators (defined loosely by impacts on juvenile plants that prevent survival to adulthood), such as agouti, peccary, tapir, and other medium-to-large mammals (Fig. 3.5,16, 17), may prevent the dominance of particular plant species or groups (Fig. 3.5, network d). The variance of this impact is buffered by predators of these herbivores, such as the jaguar (Fig 3.5, 23). Both predators and their herbivore prey are subject to parasites (Fig. 3.5, 27, 28), some of which are disease vectors. The productivity of the herbivore component of this system is both a consequence and a cause of the pattern of biodiversity at higher and lower trophic levels.

"Mobile link" species (Gilbert 1980), including pollinators, seed dispersal agents and plant defense mutualists (Fig. 3.5E), have minimal impact on energy and nutrient cycles in ecological time, but are critical to the maintenance of rare species, and thus, overall biodiversity. Like energy filters, their impact on the system is disproportional to the quantity of energy exchanged.

3.5 Conclusion

The lack of comparative or experimental data on primary and secondary production and the diversity of major functional types of plants and animals for a range of sites prevents a deeper understanding of the relations of secondary production and consumer diversity to primary production and plant diversity. Lack of data is a problem for understanding virtually all ecosystems, not just tropical rain forests. Although there are many examples of single species of plants or animals being extremely important to such ecosystem processes as primary and secondary productivity, there seems to be little scientific basis for arguing that biodiversity *per se* (in the sense of total number of species of any particular functional type) is important to ecosystem functioning. In general, the greatest proportion of species richness is composed of rare species, but ecosystem processes tend to be dominated by the most abundant species. However, some inconspicuous groups of organisms, such as parasitoids, may play extremely important roles as regulators of the energy and nutrient fluxes through the most abundant primary producers. The primary argument in favor of maintaining large numbers of species is that it is nearly impossible to predict which of the currently rare species may become the dominant species under altered environmental conditions or which act as critical but inconspicuous nutrient or energy filters. The temporal and spatial scales at which these critical interactions function are largely unknown and in need of extensive study.

References

Andow D (1983) Effect of agricultural diversity on insect populations. In: Lockeretz W (ed) Environmentally sound agriculture. Praeger, New York, pp 96-115

Coley PD, Aide TM (1988) Comparisons of herbivory and plant defenses in temperate and tropical broad-leafed forests. In: Price PW, Lewinsohn TM, Fernades GW, Benson WW (eds) Plant-animal interactions. Wiley Intersci, New York, pp 25-49

Connell JH (1971) On the role of natual enemies in preventing competitive exclusion in some marine animals and in rainforest trees. In: den Boer PJ, Gradwell GR (eds) Dynamics of populations. Cent Agric Publ, Wageningen

Darwin C (1859) On the origin of species by means of natural selection or the preservation of favored races in the struggle for life. Murray, London

Dix R, Smeins F (1967) The prairie, meadow and marsh vegetation of Nelson County, North Dakota. Can J Bot 45: 21-58

Gentry AW (ed) (1990) Four neotropical rainforests. Yale Univ Press, New Haven

Gilbert LE (1977) The role of insect-plant coevolution in the organization of ecosystems. In: Labyrie V (ed) Comportement des insectes et milieu tropique. CNRS, Paris, pp 399-413

Gilbert LE (1980) Food web organization and the conservation of neotropical diversity. In: Soulé M, Wilcox BA (eds) Conservation biology. Sinauer, Sunderland, MA, pp 11-33

Golley FB (1960) Energy dynamics of a food chain of an old-field community. Ecol Monogr 30:187-206

Hubbell SP (1973) Populations and simple food webs as energy filters. I One-species systems. Am Nat 107:194-201

Huston MA (1979) A general hypothesis of species diversity. Am Nat 113:81-101

Huston MA (1992) Biological diversity and human resources. Impact of science on society. UNESCO 166:121-130

Huston MA (1993) Biological diversity, soils, and economics. Science 262:1676-1680

Huston MA (1994) Biological diversity: The coexistence of species on changing landscapes. Cambridge Univ Press, Cambridge

Janzen DH (1970) Herbivores and the number of tree species in tropical forests. Am Nat 104:501-28

Jordan CF (1983) Productivity of tropical rain forest ecosystems and the implication for their use as future wood and energy resources. In: Golley FB (ed) Tropical rain forest ecosystems, structure and function. Ecosystems of the world, vol. 14A. Elsevier, Amsterdam, pp 117-36

Lattin JD (1990) Arthropod diversity in northwest old-growth forests. Wings 15:7-10

Lawton JH, Brown UK (1993) Redundancy in ecosystems. In: Schulze E-D, Mooney HA (eds) Biodiversity and ecosystem function. Springer, Berlin Heidelberg New York, pp 361-382

McKey D (1979) The distribution of secondary compounds within plants. In: Rosenthal GA, Janzen DH (eds) Herbivores: their interactions with secondary plant metabolites. Academic Press, New York, pp 56-133

McKey DB, Waterman PG, Mbi CN, Gartlan JS, Strusaker TT (1978) Phenolic content of vegetation in two African rain-forests: ecological implications. Science 202:61-64

Owen JG (1988) On productivity as a predictor of rodent and carnivore diversity. Ecology 69:1161-65

Simms EL (1992) Costs of plant resistance to herbivory. In Fritz RS, Simms EL (eds) Plant resistance to herbivores and pathogens. Univ Chicago Press, Chicago, pp 392-425

Terborgh J (1986) Keystone plant resources in tropical forests. In: Soulé, NE (ed) Conservation biology: the science of scarcity and diversity. Sinauer, Sunderland, MA, pp 330-355

Vitousek PM (1986) Biological invasions and ecosystem properties: Can species make a difference? In: Mooney HA. Drake JA (eds) Ecology of biological invasions of North America and Hawaii. Springer Berlin Heidelberg New York, pp 161-176

Weigert RG, Evans FC (1967) Investigations of secondary production in grasslands. In: Petrusewicz K (ed) Secondary production in terrestrial ecosystems. Inst Ecol Polish Acad, Warsaw

4 Biodiversity and Biogeochemical Cycles

Whendee L. Silver[1], Sandra Brown[2], and Ariel E. Lugo[3]

4.1 Introduction

Ever-increasing human activity across the tropical landscape inevitably results in the loss of biodiversity at some spatial scales (Wilson 1988; Whitmore and Sayer 1992). For example, it is well known that the replacement of diverse tropical forests with less species-rich systems results in the loss of genetic resources and of new, potentially useful plants and animals. Global-scale changes are also likely to have an impact on biodiversity, both directly through physiological responses to climate change, and indirectly through changes in the physical environment, ecosystem processes, and species interactions (Harte et al. 1992). To better determine the outcome of human-induced changes to tropical forests, we must understand the role of biodiversity in mediating ecosystem-level processes. This chapter examines the relationship between biodiversity and biogeochemical cycles in tropical forests. We begin by defining appropriate terms for biodiversity and biogeochemistry and then build a conceptual framework for linking species and ecosystem processes. Finally, we discuss the empirical evidence documenting the effects of changes in biodiversity on energy processing and nutrient cycling in ecosystems.

4.1.1 Definitions and Concepts

The term biodiversity has many different connotations, ranging from a strict description of species composition to the complexity of interactions between organisms and their environment at all spatial scales at which life occurs. Here,

[1] Department of Biology, Boston University, 5 Cummington Street, Boston Massachusetts 02215, USA
[2] Department of Forestry, W-503 Turner Hall, 1102 Gooddwin, University of Illinois, Urbana, Illinois 61801, USA
[3] International Institute of Tropical Forestry, USDA Forest Service, Callbox 25000, Rio Piedras, Puerto Rico 00928, USA

Ecological Studies, Vol. 122
Orians, Dirzo and Cushman (eds) Biodiversity and Ecosystem Processes in Tropical Forests
© Springer-Verlag Berlin Heidelberg 1996

we are concerned with two major aspects of biodiversity: species richness and functional diversity. Species richness (the number of species per unit area) is perhaps the most commonly measured component of biodiversity, and has received the greatest attention both from theorists and in empirical studies. Our index of species richness for this chapter is the number of trees, understory plants, and epiphytes per area of ecosys- tem, although considerably more data are available for trees than for the other life-forms.

We define functional diversity as a set of ecosystem processes that contribute to the maintenance of matter and energy flows through the ecosystem. Examples of the diversity of biogeochemical processes are nutrient capture, nutrient retention, nutrient transfer, and nutrient recapture. Within these general processes, a second order of diversity can be identified consisting of the attributes participating in each process. For example, nutrients are captured from atmospheric sources by tree canopies, and nitrogen fixation, and from soil by fine roots, and/or mycorrhizal fungi. Nutrients can also be recaptured by fine roots and/or mycorrhizal fungi following the release of nutrients during the decomposition of organic matter. Nutrient retention occurs by storage in biomass or on soil exchange sites. Nutrients are transferred by retranslocation, or synchronization of decomposition and litter inputs. Many functional processes may be achieved through multiple mechanisms (i.e., fine roots and microbes both capture nutrients from weathering products), which provide alternative pathways for the flow of nutrients. This multiplicity in nutrient cycling mechanisms may increase ecosystem resiliency to perturbations, where species or processes can be reduced or removed by disturbance events (Tilman and Downing 1994).

Interactions between species richness and functional diversity occur where species or groups of species participate directly in a given ecosystem process (i.e., epiphytes and nutrient capture). In an effort to link species with ecosystem functioning, the term functional groups has been used to group organisms by physiological, morphological, and phenological attributes to a particular ecosystem process (Vitousek and Hooper 1993). Both the functional diversity and functional group concepts are difficult to operationalize. Identifying a particular interaction between a biogeochemical process and a functional group can be difficult in nature, where organisms participate in many different ecosystem processes that occur not in isolation, but interdependently across multiple temporal and spatial scales. Furthermore, differentiating between the influence of different species on a particular biogeochemical process can be difficult, and we are limited by the precision of our analytical techniques to detect biogeochemical changes on the scale at which functional groups may operate. Moreover, species differences are likely to be reflected in the cycling of individual elements, of which more than a dozen are considered essential for plant growth, requiring complex levels of analysis to identify where connections between species and processes occur.

Defining appropriate terms for nutrient cycling is equally challenging. As the term implies, biogeochemical cycling describes a dynamic process that entails the interaction of biotic and abiotic components of the ecosystem. Traditionally, biogeochemistry dealt with large-scale (global and regional) nutrient cycles under geological, climatic, and biotic control. In this chapter, we use the same approach but at the watershed or individual stand-level. The parameters of consideration include inputs, exports, pool sizes, and rates of transfer between pools.

The soil is generally the largest repository for most nutrient elements in the ecosystem, and the direct recipient of inputs from weathering (Jenny 1980). A wide range of nutrient values have been reported for tropical forest soils, stemming at least in part from the numerous field and laboratory methods employed to estimate soil nutrient content (Silver 1994; Silver et al. 1994). Therefore, caution should be used when directly comparing data from different studies, as the effects of different field and laboratory protocols on nutrient values are unknown. Furthermore, because most soil extraction protocols have been developed for agricultural systems (Page 1982), the relationships among commonly used indices of soil nutrient availability to mature tropical forest vegetation are also poorly understood (Silver 1994).

These difficulties have led to the use of separate plant-oriented indices to estimate soil nutrient availability, or to characterize within-stand nutrient cycling (cf. Vitousek 1982). Measures include litterfall mass and litterfall nutrient content (Vitousek 1982, 1984; Vogt et al. 1986), litterfall mass and nutrient turnover, which considers decomposition processes (Vogt et al. 1986; Lugo 1992), whole-plant above- and belowground production and nutrient content (Cuevas et al. 1991; Lugo 1992), and standing stocks of biomass and nutrients (Greenland and Kowal 1960; Edwards and Grubb 1982; Jordan 1985). Information on factors such as plant tissue secondary chemistry, canopy and soil leaching, and nitrogen fixation can add considerable depth to the understanding of nutrient cycling in tropical forests, although few data are available. Also conspicuously absent are data on whole-ecosystem nutrient budgets for tropical forests which characterize the inputs to and outputs from the ecosystem. We argue that differences in inputs and outputs are the key to determining the relationship between biodiversity and biogeochemical cycling because they provide a measure of how open or closed ecosystem nutrient cycles are (Odum 1969; DeAngelis et al. 1992). Data on nutrient budgets would not only help to clarify which internal nutrient cycling indices were best representative of ecosystem fluxes, but they would establish a baseline from which to monitor the effects of changes in biodiversity or other ecosystem parameters. For the time being, our efforts are confined to crude comparisons of seemingly similar data and post-hoc evaluation of studies that were designed with other questions in mind. Nevertheless, interesting patterns emerge that highlight intriguing questions and future research possibilities.

4.2 Species Richness and Biogeochemical Cycling

High species richness and low soil fertility are two attributes that have historically been associated with tropical forest ecosystems. Over the years, considerable discussion has ensued regarding the role of nutrient cycling in the maintenance of biodiversity (Huston 1979, 1980; Tilman 1982; Gentry 1988), and more recently, the role of biodiversity in biogeochemistry (Ewel et al. 1981; Schulze and Mooney 1993; Vitousek and Hooper 1993). Nevertheless, relationships between species richness and soil fertility or nutrient cycling are poorly understood. This is due in part to the difficulties in reconciling the different scales at which populations and biogeochemistry are measured (Ehleringer and Field 1993), and in identifying ecological and physiological mechanisms that link a diverse and complex tropical forest flora to biogeochemical processes. Nutrient cycling in itself is a complex topic, and measurements are time consuming and expensive. We have only just begun to recognize the spatial and temporal heterogeneity in nutrient cycling in tropical forests, and to relate these to patterns in the composition and distribution of the vegetation (Tanner 1977; Edwards and Grubb 1982; Proctor et al. 1983, 1988; Lugo 1992). Results from these studies have stressed that nutrient cycling entails the complex interaction of species with their biogeochemical environment.

Vitousek and Hooper (1993) recently proposed three models of the relationship between species richness and biogeochemical cycling, assuming all other factors were equal. The first model (type 1) proposed a linear relationship, with each additional species added having a constant effect on ecosystem biogeochemistry. The second relationship (type 2) was asymptotic, with the addition of each new species having less and less effect on ecosystem nutrient cycles until the effect disappeared. The third model (type 3) proposed no effect of species richness on biogeochemical cycling after an initial suite of species representing the important functional groups were established (e.g., plants and decomposers). Vitousek and Hooper (1993) felt that the type-2 model would prove to be the most widespread in natural ecosystems, under the condition that species additions were initially increasing the diversity of functional groups, but later were adding more species to functional groups already present. Alternatively, biogeochemical cycling could influence the number of species occurring within functional groups. For example, if nutrients are cycled in a relatively closed system, i.e., nutrients are cycled in synchrony with the metabolism of the existing species in the stand and the rate of nutrient input equals the rate of nutrient export, then one would not expect the number of species within a functional group to increase. On the other hand, if nutrients are added, or are cycled under non steady-state conditions, then the possibility of species additions to functional groups should increase. Experimental studies identifying mechanistic links between species or functional groups and biogeochemical cycling in tropical forest are lacking.

4.3 Functional Diversity and Biogeochemistry

The absence of a mechanistic relationship between species richness and soil nutrient concentrations in tropical forests is not surprising when the role of species in biogeochemical cycling is considered. Tropical biota have little control over the rate and amount of nutrient inputs to an ecosystem, but they can control the amount and rate of nutrient capture, recapture, retention, and cycling in tropical forests. From this perspective, it is not necessarily the number of species that determines the type and rate of biogeochemical processes, but the role those species play as conduits for nutrient inputs and mitigators of nutrient losses. Most organisms are likely to contribute in some way to nutrient cycling in most ecosystems, but some species or groups of species exert especially strong influences on ecosystem structure. These controls occur at junctures or along critical pathways that regulate nutrient capture, the rate at which nutrients are cycled internally, and nutrient conservation by the ecosystem (Fig. 4.1). Such interfaces are zones of high metabolic activity, where resources are concentrated and cycled within and between the abiotic (atmosphere, hydrosphere, and soil) and biotic (plant, animal, and microbial) components of the ecosystem (Lugo and Scatena 1992; Silver et al. 1996).

Table 4.1 provides examples of key functional processes in tropical ecosystems, organisms or attributes that participate in them, and the mechanisms by which they occur. Our purpose here is not to provide an exhaustive list, but to illustrate the key process concept and build a conceptual framework. We discuss the examples in Table 4.1 to highlight the role of these organisms in regulating biogeochemical processes in tropical forests. Many of our examples are taken from ecosystems under stress or following disturbances. When ecosystems are stressed, key processes are often emphasized and hence stressed systems are useful for our purposes here (Lugo 1987). The same processes occur in less stressful environments, but may not be as easily identifiable due to the considerable multiplicity in biogeochemical processes and possible nutrient cycling pathways (e.g., nutrient capture and recapture occurring at more than one interface and by different organisms or attributes).

Disturbances, on the other hand, can be a confounding factor when characterizing nutrient cycles. Plants often respond to disturbances by altering mechanisms of nutrient uptake or release (Ewel et al. 1981; Silver and Vogt 1993). Following disturbance, the time necessary for the recovery of nutrient cycling processes will vary depending upon the nature of the disturbance and attributes of the ecosystem. For this reason, it is important to consider the age and successional state of the ecosystem when discussing biogeochemical cycles, and especially when comparing across sites (Silver et al. 1996).

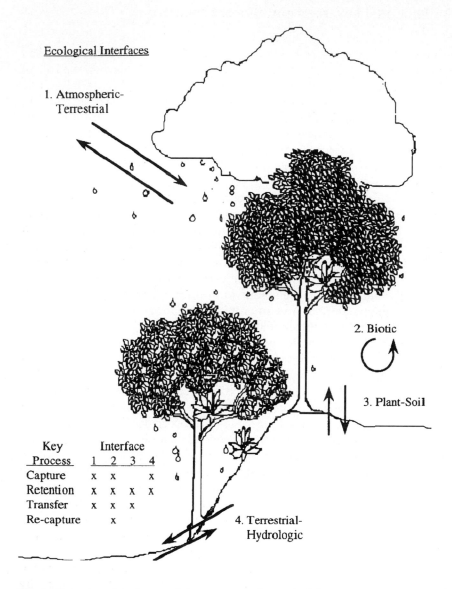

Fig. 1. Ecosystem interfaces and key processes in tropical forests. Key processes within ecological interfaces to maximiz energy flow by acquiring, retaining, and cycling nutrients in the ecosystem

Table 4.1. Examples of key processes in tropical ecosystems, the interface where they occur, the organisms or attributes that contribute to each key process, and the mechanism through which they are accomplished

Ecological interface and key process	Responsible organism(s) or attribute	Mechanisms
ATMOSPHERIC-TERRESTRIAL INTERFACE		
Nutrient capture,	Epiphytes	(1) Increase surface area exposed to atmospheric nutrient inputs
	Nitrogen-fixing organisms	(1) Fix atmospheric N in tissues
Nutrient retention	Epiphytes	(1) Immobilize nutrients in tissues and concentrate in stored water and arboreal soil; (2) Reduce the rate of water flow
Nutrient transfer	Epiphytes	(1) Cycle nutrients to terrestrial biota in litterfall; (2) Channel nutrients in stemflow and throughfall
	Nitrogen-fixing organisms	(1) Transfer N to soils and plants in throughfall, litterfall, and decay
BIOTIC INTERFACE		
Nutrient retention,	Live plant tissues	(1) Nutrient storage in tissues; (2) Produce secondary chemicals to reduce herbivory
Nutrient transfer	Live and dead plant tissues	(1) Retranslocate nutrients within tissues to minimize nutrient losses; (2) Produce litter
PLANT-SOIL INTERFACE		
Nutrient capture Nutrient recapture	Rootmats and high fine root biomass	(1) Capture nutrients prior to release to the soil
	Mycorrhizae	(1) Increase nutrient availability to plants by increasing exploitation of the rhizosphere; (2) Capture nutrients prior to release to the soil
	Symbiotic N-fixers	(1) Fix soil atmospheric N and transfer directly to plant roots
Nutrient retention	Rootmats and high fine root biomass	(1) Reduce the rate of water flow through the soil; (2) Store organic matter and nutrients in tissues
Nutrient transfer	Rootmats and high fine root biomass	(1) Transfer nutrients from soil to plants

Table 4.1. cont.

Ecological interface and key process	Responsible organism(s) or attribute	Mechanisms
	Mycorrhizae	(1) Transfer nutrients from soil to plants
	Litterfall	(1) Control decomposition rates by the production of secondary compounds; (2) Synchronize nutrient mineralization through control of litter quality and litter inputs
TERRESTRIAL-HYDROLOGIC INTERFACE		
Nutrient capture and Nutrient recapture	Fine Roots and Soil Microbes	(1) Capture nutrients from stream water and alluvial sediments
Nutrient retention	Roots	(1) Store nutrients in tissues; (2) increase drainage
	Coarse Woody Debris	(1) Reduce the rate of surface flow of water and retain litter

4.3.1 The Atmospheric-Terrestrial Interface

From a nutrient cycling perspective, the primary process accomplished by species occupying the atmospheric-terrestrial interface is the acquisition of nutrients through interception of cloud moisture, rainfall, and dry deposition, and from direct fixation from the atmosphere (Fig. 4.1). Epiphytes contribute substantially to nutrient acquisition from the atmosphere, and tropical forests, in particular montane tropical forests, are well known for their high biomass and species richness of epiphytes (Gentry and Dodson 1987; Benzing 1990). Unlike terrestrially rooted plants, epiphytes receive most of their nutrients from atmospheric sources, and thus provide a major pathway for the transfer of nutrients from atmospheric sources to the terrestrially rooted vegetation. This results from: (1) increasing nutrient capture by increased surface area in contact with cloud moisture and rainfall; (2) concentration of nutrients in epiphytic tissues (nutrient retention), stored water, and arboreal soil that is eventually cycled to the soil (nutrient transfer) in litterfall (Nadkarni and Matelson 1992); (3) channeling of nutrients to the forest floor in stemflow and throughfall (nutrient transfer); and (4) reduction of the rate at which water moves through the ecosystem by interception, which reduces the chances for nutrient loss by soil erosion (nutrient retention).

The few studies that have looked at the role of epiphytes in nutrient cycling have been conducted in tropical cloud forests (Nadkarni 1984;

Nadkarni and Matelson 1992). Nutrient availability in these ecosystems is often low due to low soil concentrations (Grubb 1977; Tanner 1977), inhibition of nutrient uptake due to extreme water logging (Lyford 1969), or periods of extended drought (Sugden and Robins 1979), and the location of the primary nutrient "substrate" aboveground. Epiphytes are particularly good at concentrating nutrients that might otherwise be lost from the ecosystem in leachate. In a cloud forest in Costa Rica, Nadkarni (1984) found that 45% of the foliage nutrient pool was stored in epiphytic biomass. Epiphytes also store nutrient rich water in cloud forests (Pocs 1982), thus further contributing to nutrient retention.

Nitrogen-fixation is another key process that occurs in the atmospheric-terrestrial interface (Table 4.1). Unlike most other nutrient elements, the major sources of N to ecosystems are precipitation and biological nitrogen-fixation from the atmosphere. Nitrogen-fixation occurs both by free-living organisms (bacteria), and by symbiotic bacterial associations with plants (Runge 1983). There are many examples of the contributions of nitrogen fixation to the N economy of both temperate and tropical ecosystems (Van Cleve et al. 1971; Vitousek et al. 1987; Montagnini and Sancho 1990). The role of nitrogen fixation as a key process becomes particularly apparent in successional communities. For example, on young lava flows in Hawaii, the invasion of a nitrogen-fixing tree quadrupled the amount of N entering the ecosystem at some sites, increasing N availability to other organisms, and altering the pattern of ecosystem development (Vitousek et al. 1987; Vitousek and Walker 1989). Selective enrichment of nitrogen fixing legumes in a mature secondary forest in Puerto Rico was estimated to increase the N immobilized in plant tissue by up to 20% (Garcia Montiel and Scatena 1994).

4.3.2 The Biotic Interface

Much of the nutrient cycling in most tropical forests occurs at the biotic and plant-soil interfaces. The division of these two interfaces is primarily a matter of scale and active participants. The biotic interface is where processes occur that involve the biota almost exclusively, whereas the plant-soil interface directly involves biotic-abiotic interactions. In reality, however, it is often difficult to separate processes in space and time. Likewise, what may occur exclusively in the biotic interface in one forest, or at one level of organization, may occur in the plant-soil interface in another. In this chapter, we consider litterfall to be part of the biotic interface and litter decomposition part of the plant-soil interface, although we recognize that this will not fit all tropical forest ecosystems, nor all possible perspective (e.g., belowground litter production, nutrient transfer from dead or material to soil microorganisms). The key biogeochemical proces biotic interface are nutrient retention and nutrient transfer (Ta¹

Nutrient retention by the biota occurs through the storage of nutrients in tissues. Tropical forests store a large quantity of nutrients in the live biomass (Jordan 1985). In addition to nutrient storage, the production of secondary chemicals may help conserve nutrients in the tissues by reducing potential losses to herbivores (McKey et al. 1978; Hobbie 1992). The relatively long life span of leaves of many tropical forest trees may also increase nutrient retention (Jordan 1991). Intimately tied to nutrient retention is within-plant nutrient transfer by retranslocation, measured as the difference between nutrients stored in live tissues and nutrients deposited in litterfall. Few studies have compared both live and senesced tissues in tropical forests, and most have relied on litterfall measures alone. In a review of tropical litterfall data, Vitousek (1984) found that the litterfall mass/litterfall P ratio was high in tropical forests, suggesting that retranslocation of P may be an important nutrient transfer pathway. Even so, because rates of litterfall tend to be high in tropical forests, particularly in lowland tropical moist forests (Lugo and Brown 1991), significant quantities of nutrients are transferred to the forest floor by this pathway.

4.3.3 The Plant-Soil Interface

Nutrients are taken up from the soil and forest floor through fine roots and eventually returned again through the decomposition of litterfall and belowground litter inputs. Thus, the key functional processes in the plant-soil interface include nutrient capture from weathering, nutrient retention on soil exchange sites and in organic matter, nutrient transfer within and between biotic and abiotic pools, and the recapture of nutrients released from decomposition. Some of the ways in which the biota can affect the rate at which nutrients move through the ecosystem are given in Table 4.1. In this section, we briefly review the effects of litter quality, including secondary chemistry, on decomposition and the role of fine roots and mycorrhizal fungi in nutrient cycling at the plant-soil interface.

Litterfall and decomposition are the primary routes of nutrient transfer in forest ecosystems. The timing, amount, and chemical composition of litterfall are important deter____nants of the rates of decomposition and nutrient release (V___ . Although considerable data are available on rates of litt___ r decomposition and nutrient release have received ____ forests. It has been proposed that nutrient release ____nts is synchronized through the timing, nutrient ____mistry of above- and belowground litter inputs ____ft 1986; Bloomfield et al. 1993). For example, fine ____ species decompose more slowly than leaf litter ____oncentrations of lignin, polyphenolics, and ____loomfield et al. 1993). Tree leaves tend to be ____ion, whereas roots are often sites of immobilization, with the combination of both presumably synchronizing nutrient release with uptake. The transfer of nutrients to the site of decay by microbes can facilitate the redistribution of nutrients across the landscape.

One of the most conspicuous features of the plant-soil interface in many humid tropical forests are rootmats and high fine root biomass in the surface soil horizons. Rootmats and surface fine roots account for a high proportion of the total root biomass in many tropical forests (Sanford 1985; Silver and Vogt 1993), but reach their best development in forests growing on highly leached, nutrient-poor soils (Klinge 1973; Klinge and Herrera 1978; Stark and Spratt 1977). In these ecosystems, surface fine roots, and especially mycorrhizal roots, take up nutrients directly from the litter, circumventing the mineral soil where nutrients could be easily lost (Went and Stark 1968; Stark 1971; Stark and Jordan 1978; Cuevas and Medina 1988). Fine roots also capture nutrients from water percolating through the rootmat (Stark and Jordan 1978). The magnitude of nutrient conservation by fine roots is difficult to quantify, but is likely to be high. In a montane forest in Puerto Rico, Silver and Vogt (1993) found that fine root mortality following disturbance significantly increased nitrification, lowered soil pH, and decreased soil cation pools.

4.3.4 The Terrestrial-Hydrologic Interface

Most nutrient export from ecosystems is through the soil. A considerable proportion of tropical forests occur on deep, highly weathered soils, with low nutrient holding capacity (Sanchez 1976). In the humid tropics, large volumes of water moving through the soil further increase the chances for nutrient losses through leaching (Radulovich and Sollins 1991) and erosion. Therefore, the terrestrial-hydrologic interface is another important juncture for nutrient cycling in tropical forests where the primary functional processes are nutrient capture, nutrient recapture, and nutrient retention.

Coarse woody debris reduces the rate of water movement in and on the soil, decreasing nutrient loss by erosion (Brady 1990), and thus contributing to ecosystem nutrient retention. Coarse woody debris also contributes to nutrient retention both through nutrients immobilized in the wood during decay (Brown and Lugo, unpubl. data) and by blocking the movement of fine litter, resulting in the accumulation of additional decay substrates.

In addition to participating in key processes in the plant-soil interface, roots also reduce nutrients losses in the terrestrial-hydrologic interface. Both fine and coarse roots can increase drainage in the soil, thereby reducing overland flow and soil erosion. Roots also aerate the soil, reducing the chances for anaerobic conditions to develop, which could reduce nutrient uptake. Nutrient uptake by fine roots, storage in root tissues, and transfer to other plant parts removes nutrients from the soil where leaching losses occur. Rootmats and coarse roots on the soil surface slow the rate of overland flow and block downslope movement of litter and nutrients.

An excellent example of nutrient cycling in the terrestrial-hydrologic interface is found in tropical floodplain forests. Because floodplain forests are often tightly coupled with surrounding forest ecosystems, they regulate the loss and capture of nutrients within a larger landscape complex. Nutrients are captured from sediments and organic material deposited during high water or flood events, and recaptured following release from upslope sources, reducing nutrient loss to stream water. Flooding can also reduce nutrient uptake by the floodplain biota, through mechanical damage to individuals, the creation of anaerobic conditions, and through losses due to leaching and down stream transport (Frangi and Lugo 1985).

4.4 Evidence from Experimental Studies

To determine the effects of species richness and functional diversity on nutrient cycling requires experimental studies where all factors except the number of species or a particular process are held constant. However, because experimental manipulations are essentially a disturbance to the system, it is difficult to separate disturbance effects from the effects of changes in species or processes. Likewise, we lack good estimates of the time required to reestablish nutrient cycles following a disturbance event, although it will probably vary between ecosystems and disturbances. There are very few experimental studies in tropical forests with which the effects of species richness and functional diversity can be addressed. In the absence of studies designed with this question in mind, we report on two studies that provide insights into the topic: (1) a study by Lugo (1992) that compared nutrient cycling processes between plantations of different ages and species compositions (lower species richness) and paired secondary forests (higher species richness) in Puerto Rico; and (2) a study by Ewel et al. (1991) on the effects of plant diversity, manipulated as a treatment, on nutrient cycling processes in young successional ecosystems in Costa Rica.

4.4.1 Plantations Versus Natural Forests

The study by Lugo (1992) compared four plantations; two pine (*Pinus caribaea* that were 4 and 18.5 years old) and two mahogany (*Swietenia macrophylla* that were 17 and 49 years old) with paired, adjacent secondary forests of similar ages, growing under similar edaphic and climatic conditions. The only prior management in these plantations was some selective thinning on a few occasions in the older pine and mahogany plantations. All sites are in the subtropical wet forest life zone (sensu Holdridge 1967) growing on

Table 4.2. Species richness in paired plantations and secondary forests of similar ages. (Lugo 1992)

Site	Number of trees species	Number of understory species
1. 4-year-old pine plantation	6	17
Secondary forest	17	23
2. 18.5-year-old pine plantation	15	16
Secondary forest	16	26
3. 17-year-old mahogany plantation	15	25
Secondary forest	24	31
4. 49-year-old mahogany plantation	16	30
Secondary forest	37	38

ultisols or oxisols. The numbers of tree and understory plant species were higher in the secondary forest than in the plantation pair, and the numbers increased with increasing age of the plantation-secondary forest pair as shown in Table 4.2 (Lugo 1992).

All plantations had more aboveground biomass (40-171 Mg/ha) but less belowground biomass (2.8-19.8 Mg/ha) than the paired secondary forests (aboveground biomass of 32-109 Mg/ha and belowground biomass of 10.9-21.3 Mg/ha). Aboveground production was also higher in the plantations than in the paired secondary forests (11.4-19.1 Mg ha^{-1} yr^{-1} and 7.9-12.3 Mg ha^{-1} yr^{-1}, respectively), caused mainly by higher litterfall rates. Remeasurement 7 years later of both the 4-year-old pine plantation and its secondary forest pair showed that the total above- and belowground production of the two forests was the same (about 19 Mg ha^{-1} yr^{-1}) (Cuevas et al. 1991). The main difference was in the allocation of organic matter production. Root production was very low in the plantation, whereas root production in the secondary forest was almost half the total production. We expect that similar patterns exist for the other plantation pairs based on the distribution of above- and belowground biomass and aboveground production.

We believe that the differences in organic matter production and allocation are a response to different strategies represented in the forest species assemblages to acquiring and retaining nutrients (Fig. 4.1, Table 4.1). Here, we describe some biogeochemical processes from Lugo's study that illustrate how different mechanisms perform the same functional role. These mechanisms appear to arise because of functional differences between species (e.g., gymnosperms versus angiosperms) rather than absolute number of species, although the relatively young age of the forests could also be a factor. Because of the scope of Lugo's study, we discuss only the biotic and plant-soil interfaces. Furthermore, although every attempt was made to select stands on the same soil type, results of the study did show differences among pairs. However, these differences were small and there did not appear to be significant nutrient limitations at any of the sites.

Despite their lower species richness, plantations produced more aboveground biomass per unit of nutrient uptake than secondary forests did, with the exception of the young pine plantation that was as efficient as its paired secondary forest. The less species-rich plantations generally contained significantly larger pools of N and P, particularly in aboveground vegetation and forest floor. Soil N and P pools were higher in the pine plantations than in the paired secondary forests, whereas the trend was reversed in the mahogany plantations and secondary forest pairs. Part of this was most likely caused by the age differences as well as initial soil conditions (Lugo 1992). Trends for K, which is more mobile, were less clear.

The differences in biogeochemical cycling that most likely cause the higher stand-level efficiency in plantations are summarized in Table 4.3. Even though the plantations have lower species richness than the secondary forests, they appear to have slightly higher functional diversity. The secondary forests exhibited two nutrient cycling mechanisms that are

Table 4.3. Generalized differences in mechanisms for acquiring and retaining nutrients (N, P, and K) in plantations and secondary forest in Puerto Rico. Exceptions to these trends often occurred in one plantation-secondary forest pair. Mechanisms are grouped by biogeochemical processes within the biotic and plant-soil interfaces (Table 4.1). (Based on data in Lugo 1992)

Key process	Plantation	Secondary forest
Nutrient capture	More mycorrhizal connections[a]	Greater root mass and number, and deeper rooting depths[*]
Nutrient retention	1. Higher aboveground biomass and N and P pools	1. Higher root nutrient pools and more root biomass
	2. Greater quantity and rate of accumulation of nutrients in forest floor	2. Low litterfall mass
	3. More nutrient immobilization in the forest floor	
Nutrient transfer	1. Higher rate of nutrient retranslocation	1. Greater nutrient concentration in litterfall and thus greater return in litterfall
	2. Higher litterfall mass	2. Greater mix of litter quality, but lower nutrient-use efficiency[*]
		3. Faster rates of litter decomposition and nutrient mineralization

[*] Function of plant species richness
[a] Not reported in original study (Lugo, pers. Comm.)

likely to be a direct function of increased species richness, i.e., greater rooting depth and a greater mixture of litter qualities. Based on the results of Lugo's work, we conclude that for relatively young plantations growing on relatively nutrient-rich soils, decreased species richness does not appear to result in significant nutrient deficiencies or decreased biomass production. In the study by Lugo, however, plantations fostered relatively species-rich understory and tree communities, all of which participate in nutrient cycling. In fact, true perennial monocultures that do not occur naturally are very difficult to maintain in tropical forests sites, suggesting that succession tends to fill in both the number of functional groups, and increase the size of functional groups where options for nutrient acquisition are available.

4.4.2 Experimental Manipulation of Species Composition

Ewel et al. (1991) studied the influence of five treatments of species richness on ecosystem structure and functioning in a tropical moist forest in Costa Rica. The five treatments were bare soil (no plants); a successive planting of two short-lived monocultures ending with a tree plantation of *Cordia alliodora*; a natural succession; an imitation of natural succession that contained the same number of species and growth forms as the natural succession with no overlap in species composition and no native species; and an enriched succession where propagules were added to the natural species assemblage. At the end of 5 years, the natural succession and the imitation contained about 100 species each, and the enriched succession contained about 25% more species.

Soil nutrient pools were greater in the more diverse treatments, which was attributed to more effective nutrient retention and maintenance of soil processes favorable for plant production (Ewel et al. 1991). Nutrient depletion was greater in sites that were less fertile to begin with, and the monocultures were more severely affected by depletion than the diverse treatments. The short-lived monocultures experienced the greatest nutrient losses, whereas the tree plantation exhibited only minimal nutrient losses (Ewel et al. 1991). In the framework of our model of functional diversity, the tree plantations are probably more functionally diverse than the short-lived monocultures, and thus are better able to retain nutrients in the ecosystem. Functional diversity is probably greatest in the plant-soil interface, where the greater fine root biomass and microbial populations enhance nutrient capture, recapture, and transfer, and where the greater diversity of litter inputs from different forest strata contribute to nutrient cycling. No differences occurred in soil fertility between the diverse treatments, suggesting that the additional species did not increase the functional diversity in these already species-rich systems.

Results from these studies provide ample fodder for future research to examine the effects of biodiversity on biogeochemical cycling. Based on these studies, we would hypothesize that at least a minimum level of species richness is necessary to maintain nutrient retention and biomass production in tropical ecosystems, and that the characteristics of those species (e.g., gymnosperm versus angiosperm, annual versus perennial) can have a stronger influence on biogeochemical processes than just the number of species occurring on a site. These studies also illustrate that there are often multiple possible pathways for the flow of nutrients and matter in tropical ecosystems, which has significant implications for ecosystem management. Management that maintains the functional diversity of the ecosystem should, in theory, result in the maintenance of biogeochemical cycling and productivity (Silver at al. 1996).

4.5 Conclusions

Tropical deforestation and global-scale atmospheric changes are altering, or have the potential to alter, biodiversity in tropical regions. The vast majority of attention has been focused on the loss of species, and the consequent loss of genetic resources and human services. Does the loss of species also result in alterations in biogeochemical cycling? The few data available suggest that the role species play in biogeochemical cycles may be more important than the number of species on a site, particularly in relatively diverse ecosystems. When the focus is shifted from the diversity of species to the diversity of ecosystem functional processes, the number of species becomes less important as long as biogeochemical processes are maintained.

Clearly more research is needed to better determine the role of biodiversity in ecosystem functioning in tropical forests. We stress the importance of functional diversity, and present the key functions concept as a framework for understanding the role of organisms in biogeochemical cycles, and for designing manipulative experiments to determine the interactions between organisms and their environment. We also suggest that ecosystems under stress, while often discounted as 'different', may prove to be the best testing grounds for studying key functions and understanding the role of biodiversity in biogeochemical cycles.

References

Ashton PS (1977) A contribution of rain forest research to evolutionary theory. Ann Mo Bot Gard 64:694-705

Benzing DH (1990) Vascular epiphytes. Cambridge Univ Press, Cambridge

Bloomfield J, Vogt KA, Vogt DJ (1993) Decay rate and substrate quality of fine roots and foliage of two tropical tree species in the Luquillo Experimental Forest, Puerto Rico. Plant Soil 150: 233-245

Brady NC (1990) The nature and properties of soils. Macmillan, New York

Cuevas E, Medina E (1988) Nutrient dynamics within Amazonian forests 2. Fine root growth nutrient availability and leaf litter decomposition Oecologia 76: 222-235

Cuevas E, Brown S, Lugo AE (1991) Above-and belowground organic matter storage and production in a tropical pine plantation and a paired broadleaf secondary forest. Plant Soil 135: 257-268

DeAngelis DL, Mulholland PJ, Palumbo AV, Steinman AD, Huston MA, Elwood JW (1992) Nutrient dynamics and food-web stability. Annu Rev Ecol Syst 20:71-95

Edwards PJ, Grubb PJ (1982) Studies of mineral cycling in a montane rain forest in New Guinea IV Soil characteristics and the division of mineral elements between the vegetation and soil. J Ecol 70:649-666

Ehleringer JR, Field CB (eds) (1993) Scaling physiological processes. Academic Press, San Diego

Ewel JJ, Berish C, Brown B, Price N, Raich J (1981) Slash and burn impacts on a Costa Rican wet forest site. Ecology 62:816-829

Ewel JJ, Mazzarino MJ, Berish CW (1991) Tropical soil fertility changes under monocultures and successional communities of different structure. Ecol Appl 1:289-302

Frangi JL, Lugo AE (1985) Ecosystem dynamics of a subtropical floodplain forest. Ecol Monogr 55:351-369

Garcia-Montiel DC, Scatena FN (1994) The effect of human activity on the structure and composition of a tropical forest in Puerto Rico. For Ecol Manage 63:57-78

Gentry AH (1988) Changes in plant community diversity and floristic composition on environmental and geographical gradients. Ann Mo Bot Gard 75:1-34.

Gentry AH Dodson C (1987) Contribution of nontrees to species richness of a tropical rain forest. Biotropica 19:149-156

Greenland DJ, Kowal JML (1960) Nutrient content of the moist tropical forest of Ghana. Plant Soil 12:154-174

Grubb PJ (1977) Control of forest growth and distribution on wet tropical mountains: with special reference to mineral nutrition. Annu Rev Ecol Syst 8:83-107

Harte J, Torn M, Jensen D (1992) The nature and consequences of indirect linkages between climate change and biological diversity. In: Peters RL, Lovejoy TJ (eds) Global warming and biological diversity. Yale Univ Press, New Haven, pp 325-343

Hobbie SE (1992) Effects of plant species on nutrient cycling. Trends Ecol Evol 7:336-339

Holdridge LR (1967) Life zone ecology. Tropical Science Center, San José, Costa Rica

Huston M (1979) A general hypothesis of species diversity. Am Nat 113:81-101

Huston M (1980) Soil nutrients and tree species richness in Costa Rican forests. J Biogeogr 7:147-157

Jenny H (1980) The soil resource. Springer, Berlin Heidelberg New York.

Johnston MH (1992) Soil-vegetation relationships in a tabonuco forest community in the Luquillo Mountains of Puerto Rico. J Trop Ecol 8:253-263

Jordan CF (1985) Nutrient cycling in tropical forest ecosystems. Wiley, New York

Jordan CF (1991) Nutrient cycling processes and tropical forest management. In: Gómez-Pompa A, Whitmore TC, Hadley M (eds) Rain forest regeneration and management. MAB series, vol 6. Parthenon, New Jersey, pp 159-180

Klinge H (1973) Root mass estimation in lowland tropical rain forests of central Amazonia, Brazil. 1. Fine root masses of a pale yellow latosol and a giant humus podzol. Trop Ecol 14:29-38

Klinge H (1975) Root mass estimation in lowland tropical rain forests of central Amazonia, Brazil. 3 Nutrients in fine roots from giant humus podzols. Trop Ecol 16:28-38

Klinge H, Herrera H (1978) Biomass studies in Amazonia Caatinga forest in southern Venezuela 1. Standing crop of composite root mass in selected stands. Trop Ecol 19:93-101

Lawton JH, Brown VK (1993) Redundancy in ecosystems. In: Schulze ED, Mooney HA (eds) Biodiversity and ecosystem function. Springer, Berlin Heidelberg New York, pp 255-270

Lugo AE (1987) Stress and ecosystems. In: Thorp JH, Gibbons JW (eds) Energy and environmental stress in aquatic ecosystems. DOE Symp Ser (Conf-771114) Nat Tech Inf Serv Va, pp 62-101

Lugo AE (1992) Comparison of tropical tree plantations with secondary forests of similar age. Ecol Monogr 62:1-41

Lugo AE, Brown S (1981) Tropical lands: popular misconceptions. Mazingara 5:10-19

Lugo AE, Brown S (1991) Comparing tropical and temperate forests. In: Cole, JC, Lovett GM, Findlay SEG (eds) Comparative analysis of ecosystems: patterns, mechanisms, and theories. Springer, Berlin Heidelberg New York, pp 319-330

Lugo AE, Scatena FN (1992) Epiphytes and climate change research in the Caribbean: a proposal. Selbyana 13:123-130

Lyford WH (1969) The ecology of an elfin forest in Puerto Rico. 7 Soil, root, and earthworm relationships. J Arnold Arbor 50:210-224

Mckey DP, Waterman G, Gartlan JS, Struhsaker TT (1978) Phenolic content of vegetation in two African rain forests: ecological implications. Science 202:61-64

Montagnini F, Sancho F (1990) Impacts of native trees on tropical soils: a study in the Atlantic lowlands of Costa Rica. Ambio 19:386-390

Nadkarni NM (1984) Epiphyte biomass and nutrient capital of a neotropical elfin forest. Biotropica 16:249-256

Nadkarni NM, Matelson TJ (1992) Biomass and nutrient dynamics of epiphytic litterfall in a neotropical montane forest, Costa Rica. Biotropica 24:24-30

Odum EP (1969) The strategy of ecosystem development. Science 164: 262-270

Page AL (ed) (1982) Part 2 Chemical and microbiological properties. Am Soc Agron, Madison, Wisconsin

Pocs T (1982) Tropical forest bryophytes. In Smith AJE (ed) Bryophyte ecology. Chapman and Hall, London, pp 59-104

Proctor J, Anderson JM, Vallack HW (1983) Comparative studies on forests, soils, and litterfall at four altitudes on Gunung Mulu, Sarawak. Malays For 46:60-76

Proctor J, Lee YF, Langley AM, Munro WRC, Nelson T (1988) Ecological studies on Gunung Silam, a small ultrabasic mountain in Sabah, Malaysia. I. Environment, forest structure, and floristics. J Ecol 76:320-340

Radulovich R, Sollins P (1991) Nitrogen and phosphorus leaching in zero-tension drainage from a humid tropical soil. Biotropica 23:231-232

Runge M (1983) Physiology and ecology of nitrogen nutrition. In: Lange OL, Nobel PS, Osmond CB, Ziegler H (eds) Physiological plant ecology III. Responses to the chemical and biological environment. Springer, Berlin Heidelberg New York, pp 163-200

Sanchez P A (1976) Properties and management of soils in the tropics. Wiley, New York

Sanford RL (1985) Root ecology of mature and successional Amazon forests. PhD Diss, Univ Calif, Berkeley

Silver WL (1994) Is nutrient availability related to plant nutrient use in humid tropical forests? Oecologia 98:336-343

Silver WL, Vogt KA (1993) Fine root dynamics following single and multiple disturbances in a subtropical wet forest ecosystem. J Ecol 81:729-738

Silver WL, Scatena FN, Johnson AH, Siccama TG, Sanchez MJ (1994) Nutrient availability in a montane rain forest in Puerto Rico: spatial patterns and methodological considerations. Plant Soil 164:129-145

Silver WL, Browns, Lugo AZ (1996) Effects of changes in biodiversity on ecosystem functions in tropical forests. Con Bio 10:17-24

Stark N (1971) Nutrient cycling pathways and litter fungi. BioScience 22:355-360

Stark N, Jordan CF (1978) Nutrient retention by the root mat of an Amazonian rain forest. Ecology 59:434-437

Stark N, Spratt M (1977) Root biomass and nutrient storage in rain forest oxisols near San Carlos de Río Negro. Trop Ecol 18:1-9

Sugden AM, Robins RJ (1979) Aspects of the ecology of vascular epiphytes in Colombian cloud forests. 1. The distribution of epiphytic flora. Biotropica 11:173-188

Swift MJ (1986) Report of the third workshop on the decade of the tropics. Tropical Soil Biology and Fertility Programme. Biol Int Spec Issue, pp 13. 68

Swift MJ, Sanchez PA (1984) Biological management of tropical fertility for sustained productivity. Nat Res 20:2-10

Tanner EVJ (1977) Four montane rain forests of Jamaica: a quantitative characterization of the floristics, the soils and the foliar mineral levels, and a discussion of the interrelations. J Ecol 65:883-918

Tilman GD (1982) Resource competition and community structure. Princeton Univ Press, New Jersey

Tilman GD, Downing JA (1994) Biodiversity and stability in grasslands. Nature 367:363-365

Van Cleve K, Viereck LA, Schlentner RL (1971) Accumulation of nitrogen in alder (*Alnus*) ecosystems near Fairbanks, Alaska. Arct Alp Res 3:101-114

Vitousek PM (1982) Nutrient cycling and nutrient use efficiency. Am Nat 119:553-572

Vitousek PM (1984) Litterfall, nutrient cycling and nutrient limitation in tropical forests. Ecology 65:285-298

Vitousek PM, Hooper DU (1993) Biological diversity and terrestrial ecosystem biogeochemistry. In Schulze ED, Mooney HA (eds) Biodiversity and ecosystem function. Springer, Berlin Heidelberg New York, pp 3-14

Vitousek PM, Sanford RL (1986) Nutrient cycling in moist tropical forest. Annu Rev Ecol Syst 17:137-167

Vitousek PM, Walker LR (1989) Biological invasion by *Myrica faya* in Hawaii: plant demography, nitrogen fixation, ecosystem effects. Ecol Monogr 59:247-265

Vitousek PM, Walker LR, Whittaker LD, Mueller-Dombois D, Matson PA (1987) Biological invasion by *Myrica faya* alters ecosystem development in Hawaii. Science 238:802-804

Vogt KA, Grier CC, Vogt DJ (1986) Production, turnover, and nutrient dynamics of above- and belowground detritus of world forests. Adv Ecol Res 15:303-377

Went FW, Stark N (1968) Mycorrhiza. BioScience 18:1035-1039

Whitmore TC, Sayer JA (eds) (1992) Tropical deforestation and species extinction. Chapman and Hall, London

Wilson EO (1988) Biodiversity. Nat Acad Press, Washington DC

Wright JS (1992) Seasonal drought, soil fertility and the species diversity of tropical forest plant communities. Trends Ecol Evol 7:260-263

5 Microbial Diversity and Tropical Forest Functioning[1]

D. Jean Lodge[2], David L. Hawksworth[3] and Barbara J. Ritchie[3]

5.1 Introduction

Fungi and bacteria control many of the vital processes on which the very maintenance and survival of tropical forests depend (Hawksworth and Colwell 1992). An overview of the role of microorganisms in ecosystem functioning as a whole has already been presented (Allsopp et al. 1995). Here, our goal is to identify the functional attributes of microorganisms in tropical forests, and to identify those processes that are most likely to be sensitive to losses of diversity, especially in the face of disturbance or broad environmental changes. In some cases, microbes may influence ecosystem processes indirectly by altering the diversity of other organisms. We also point out where research is needed to determine what effects losses of microbial diversity might have on tropical forest diversity, stability, and regeneration.

5.2 The Knowledge Base

Remarkably little has been published on the numerous and often crucial ways in which fungi and bacteria influence tropical forest ecosystems. Only 96 references for fungi or bacteria in tropical forests were found by searching BIOSIS entries since 1963 as compared to 2411 references for temperate forests[4]. Most of the tropical publications related to diseases in

[1] The Forest Products Laboratory, USDA Forest Service, Palmer, Puerto Rico, is maintained in cooperation with the University of Wisconsin. This article was written and prepared by a US government employee on official time; it is therefore in the public domain and not subject to copyright.

[2] Center for Forset Mycology Research, Forest Products Laboratory, USDA Forest Service, P.O. Box B, Palmer, Puerto Rico 00721, USA

[3] International Mycological Institute, Bakeham Lane, Egham, Surrey TW20 9TY, UK

[4] Descriptors of "temperate", "coniferous", "northern", "southern", and "montane" were all included in the search for "temperate" literature

Ecological Studies, Vol. 122
Orians, Dirzo and Cushman (eds) Biodiversity and Ecosystem Processes in Tropical Forests
© Springer-Verlag Berlin Heidelberg 1996

forest plantations or descriptions of novel taxa. The few research and field observations relating the microbiota to ecosystem-level processes in tropical forests have often been of a preliminary nature, and studies of how microbial diversity per se affects ecosystem functioning are practically non-existent. Although we have tried to limit our discussion to examples from tropical forests, we have been forced at times to use examples of important processes mediated by fungi that have been studied only in temperate forests. The extrapolation of data from temperate to tropical forests may not always be valid, and research will be needed to fill this gap in our understanding of how tropical forests work.

Fungi and bacteria, apart from those causing diseases in humans, their domesticates, and crop plants, are more poorly known than are insects. The numbers of microbial species can be estimated only by extrapolation, but confidence increases where independent approaches yield comparable results. There is a growing consensus that there are between 1 and 1.5 million species of fungi on Earth (Hawksworth 1991, 1993; Hammond 1992; Rossman 1994), and perhaps as many as 3 million bacteria, most of which are as yet unculturable (Trüper 1992). Even though such large figures for the bacteria correctly meet with some skepticism, two conclusions appear inescapable: (1) that at least 90 to 95% of the Earth's microbiota has yet to be described; and (2) the microbial biota constitutes at least 15% of all species on Earth, compared with less than 3 and 0.5% for plants and vertebrates, respectively (Hammond 1992). Further, little is known of the biology, genetics, ecology, host range, biochemistry, and distribution of perhaps 40% of the 72 000 fungi currently named (Hanksworth et al. 1995).

5.3 Food Chains

In tropical forests all macroorganisms and invertebrates depend to some extent on the microbiota (Table 5.1). For example, various bacteria and fungi are involved at six points in the web supporting termites, and are critical to their survival (Price 1988). Wood- and leaf-eating insects require gut microbes to digest the lignin and cellulose in these materials, utilizing fungal enzymes in particular (Martin 1991). Fungi may be an especially important food base for some invertebrates in tropical forests, because fungal nutrient concentrations as well as fungal biomass production are relatively high (Lodge 1987b, 1993). In a lowland rainforest in Sulawesi, of 1250 beetle species caught in a 44-week trapping period, about 40 % fed directly on fungal fruit bodies or fungal mycelium in wood (Hammond 1990). Lichens serve directly as food or camouflage for a wide variety of invertebrates, and also some vertebrates in the tropics (Seaward 1988). Maintenance of fungal (including lichen) populations is thus of major importance to the conservation of a significant proportion of the insects in tropical forests.

Table 5.1. Key ecosystem processes mediated by microorganisms in tropical forests

Interface key process	Mechanism	Organism and group
Atmosphere/biotic nutrient cycling	N-fixation: autotrophic	Cyanobacteria (blue-green algae) on plants, soil, and other surfaces: 14 genera
	N-fixation: heterotrophic	On surfaces, in soil, wood, and on root surfaces - Yeasts: *Rhodotorula*, *Saccharomyces* and *Pullularia* - Bacteria: aerobic including associative N-fixers on roots *Azotobacter*, *Beijerinckia*, *Spirillum*, *Derxia*, *Azomonas* - Facultative anaerobes *Aerobacter*, *Pseudomonas*, *Bacillus*, *Achromobacter* - Anaerobic *Clostridium*, *Desulfovibrio*, *Methanobacterium*, *Rhodospirillum*, *Chromatium*, *Chlorobium*, *Rhodopseudomonas*, *Rhodomicrobium*
	N-fixation: symbiotic	With fungi, lichens formed with cyanobacteria With plant roots - Bacteria *Rhizobium* - Actinomycetes *Frankia* In animal guts Bacteria: *Enterobacter*, *Klebsiella*
	Trace gas emissions	Methanogenic bacteria Sulphur bacteria: *Thiobacillus*, *Desulfotomaculum*, *Desulfovibrio*, and many other soil organisms
	Denitrification	Bacteria: *Nitrosomonas* and 60+ other genera
Carbon cycling	Photosynthesis	Algae, cyanobacteria; independently and in lichens
	Respiration (CO_2)	All microorganisms
	Decomposition	Fungi, bacteria and actinomycetes

Table 5.1. cont.

Interface key process	Mechanism	Organism and group
Biotic/biotic interfaces		
Nutrient and carbon cycling	Nutrients and energy transfers	Plant-to-plant nutrient and carbohydrate transfers by mycorrhizal fungal "pipelines" (may be interspecific): - VA-mycorrhizal fungi (Endogonales) - Ectomycorrhizal fungi (Basidiomycotina and Ascomycotina) Host plant to achlorophyllus, parasitic or "epiphytic" higher plant transfers of nutrients/carbohydrates by: - Mycorrhizal fungi of some Bermaniaceae - Monotrapoid mycorrhizal fungi (Basidiomycotina) - Orchid mycorrhizal fungi (Basidiomycotina) particularly those associated with "epiphytic" spp.
Secondary production	Mutualisms for secondary consumption	Fungal gardens: the sole or primary food source of *Atta* and *Cyphomyrmex* ants, Old World termites, and many beetles
Secondary consumption	Microbivory	Slime molds eaten by beetles; fungi eaten by nematodes, mites, beetles, aquatic insects, millipedes, earthworms, snails, shrimp, fungus gnats, collembola, Homoptera (Dirbidae and Achilidae); bacteria eaten by nematodes, protozoans, earthworms and slime molds
Regulation of plant populations	Infection by pathogens	Bacteria, viruses, mycoplasmas, and fungi affect plant populations and dispersion. Chytridiomycota and Oomycota on algae
	Defence against herbivory and diseases?	Endophytic fungi (Ascomycotina): Non-pathogenic endophytic and ectomycorrhizal fungi of temperate plants protect their host, but their role in tropical forests is unknown

Table 5.1. cont.

Interface key process	Mechanism	Organism and group
Regulation of primary production	Competition for nutrients	Fungi and bacteria in soil, wood and litter can compete with plants for limiting nutrients, reducing production
Regulation of secondary consumers	Density-dependent disease agents	Arboviruses of birds and reptiles, viral agents and bacterial diseases of insects (e. g., *Bacillus thuringensis*); fungi killing insects, e.g., *Beauvaria Cordyceps* and *Hypocrella*; Entomophthorales on nematodes, amebae and insects
	Density-dependent disease agents	Fungi, especially Entomophthorales and *Fusarium* on mosquitoes and shrimp
	Density-dependent predators	Nematode-trapping fungi, including wood-rotting basidiomycetes; slime molds prey on other microorganisms.
Biotic/soil interfaces Nutrient cycling	Decomposition, nutrient mineralization	Fungi, bacteria and actinomycetes
Nutrient and water cycling	Mutualisms for nutrient uptake	Mycorrhizal fungi: VAM fungi (Endogonaceae); ectomycorrhizal and ericaceous basidiomycetes and ascomycetes
Soil/hydrologic interfaces Nutrient cycling	Leaching losses of N	Bacteria, actinomycetes, and fungi convert ammonium and organic N to nitrate, which is more susceptible to leaching in most tropical soils
	Weathering of rock	Algae, lichens, and chemolithotrophic bacteria, including sulfur-oxidizers, nitrifiers, and misc. heterotrophs
	Nutrient immobilization/ prevention of leaching	Fungi, bacteria and actinomycetes in soil and litter incorporate nutrients, preventing leaching losses during the rainy season

The degree to which a loss in microbial diversity from the food web of a tropical forest would affect ecosystem processes may depend on whether the microbivorous animal species or guild plays a keystone function, and the specificity of the animal-microbe interactions. Specificity is more common among symbiotic relationships, such as in fungus-gardening ants and termites (Collins et al. 1983; Wood and Thomas 1989; Cherrett et al. 1989; Giavelli and Bodini 1990), termite gut microbiota, and some of the wood-boring beetles that cultivate fungi in their galleries (Wheeler and Blackwell 1984). Anderson and Swift (1983) found that differences in decomposition rates between study sites was influenced significantly by the presence of termites. Non-symbiotic invertebrates involved in the detrital food web can have distinct feeding preferences for certain fungi or litter colonized by them, while leaving other distasteful or toxic species entirely untouched. Such preferences among invertebrates have been found to influence the rates of decomposition and nitrogen mineralization from leaf litter in temperate forests (Newell 1984) and streams (Arsuffi and Suberkropp 1984).

5.4 Pathogens

5.4.1 Control of Herbivores by Pathogens

The rates of defoliation by herbivorous insects and the losses of plant photosynthates to sucking insects in tropical forests are moderated by insect pathogens and parasitoid insects (Barbosa et al. 1991; Table 5.1). Agrochemical companies search tropical forests for entomopathogens because of the high species richness (Martin and Travers 1989; Jun et al. 1991). According to Hywel-Jones (1993), there are at least 1.5 to 13.5 million undescribed fungi that infect insects. Most parasitic fungi are host-specific, so a loss in their diversity could result in greater defoliation. Bacterial pathogens are also important in controlling infestations of caterpillars and other insects in tropical forests (Martin and Travers 1989).

5.4.2 Pathogens as a Source of Disturbance

Pathogens have the capacity to exert dramatic effects on the flora and fauna, and to convert forests to other vegetation types (Haldane 1927; Weste 1986; Castello et al. 1995). Such conversions to a different forest community or vegetation type are expected to have effects on ecosystem processes similar to conversions resulting from anthropogenic disturbances. Phenomena of this severity have only rarely been witnessed in undisturbed, natural tropical forests, which, because of their typically diverse mix of tree species, may be somewhat less vulnerable to single pathogens than temperate ones with a few dominant tree species. The devastation wrought by native *Armillaria, Fomes,* and *Ganoderma* species in tropical tea, rubber, and cacao plantations (Leach 1939; Anon. 1946; Rishbeth 1955, 1980; Pichel 1956; Fox 1970) point to the potential for epidemics to occur in natural forests if their species diversity were drastically reduced. Some natural tropical forests, however, are dominated by one or a few tree species or genera, and may be particularly susceptible to large-scale disturbances caused by pathogens. For example, oil palm forests (*Elaeis guineensis*) in Malaysia and Indonesia have enlarging gaps associated with the spread of *Ganoderma* root-rot fungi (Turner 1991). Mortality caused by *Phellinus noxius* similarly spreads through forests where the density of *Araucaria cunninghamii* (Queensland hoop pine) in Australia is high enough to permit root-to-root contacts (Bolland 1984).

Anthropogenic disturbances such as partial cutting can increase the incidence and severity of diseases that promote forest gap formation in tropical as well as in temperate forests (Castello et al. 1995). For example, *Armillaria luteobubalina* is pathogenic on many canopy *Eucalyptus* and understory species in Western Australia (Kile 1981); selective cutting has greatly increased the incidence and mortality from this disease by leaving trees in close proximity to inoculum bases of the freshly cut stumps (Edgar et al. 1976; Kellas et al. 1987; Castello et al. 1995). Pathogenic *Armillaria* species become more aggressive by producing toxins that kill healthy trees if given access to a large food base, such as a cut stump (Leach 1939; Redfern 1975; Cook 1977). Similarly, logging, in combination with stress from wildfires and high rainfall, has increased the severity of *Phytophthora cinnamomi* root rot in eastern Victoria, Australia (Fagg et al. 1986; Castello et al. 1995). Logging favors this disease by increasing soil temperatures and increasing soil moisture via reduced transpiration (Weste and Marks 1987). As a result, the proportions of resistant and susceptible tree species have changed, some rare species are endangered, overall tree density in diseased sites decreased by 43% over 10 years whereas density increased by 10% in pathogen-free sites over the same period, and forest structure changed from an open canopy with a sclerophyllous understory to a forest with large gaps dominated by sedges, grasses, and leguminous plants (Weste 1986). In South Queensland, Australia, soil bacteria that normally supress *P. cinnamomi* in the rain forest are lost following severe human disturbances because they grow poorly at the low soil pH that results from increased leaching of cations (Baker and Cook 1974; Broadbent and Baker 1974).

5.4.3 Effect of Pathogens on Patterns of Tree Dispersion

Extreme clumping of tree species may alter nutrient mineralization from litter and nutrient retention of soils, so the effect of pathogens in reducing clumping can be significant for ecosystem processes. In Costa Rica, Ewel et al. (1991) found lower total nitrogen content, phosphorus sorption capacity, and extractable calcium and other base cations in soils in monoculture and low diversity plots versus plots with high diversity. In temperate forests, leaf litter mixtures often have significantly faster rates of nitrogen mineralization than pure decomposition bags of the component species owing to differences in the decomposer communities (Ineson et al. 1982; Blair et al. 1990). Similarly, Burghouts et al. (1994) found lower rates of leaf litter decomposition in dipterocarp assemblages where species richness had been reduced by logging as compared to assemblages with greater species richness in Malaysia. Nutrient retention by ecosystems has been found to increase with plant species diversity in the range of one to ten species, and especially with diversity of functional groups that have different litter C/N ratios, but diversity of rooting patterns may also be involved (Hooper and Vitousek 1992; Vitousek and Hooper 1993).

Certain tropical tree species are known to change nutrient cycling characteristics beneath their canopies. For example, soils beneath nitrogen-fixing trees have lower pH values as a result of increased microbial denitrification (e.g., *Pentaclethra macroloba* in Costa Rica; Parker 1994). Clumping of such trees could have a disproportionate influence on the fate of nutrients in the ecosystem because of nonlinear responses to increased nutrient concentrations (Lodge et al. 1994).

Janzen (1970) and Connell (1971) independently proposed that escape from natural enemies that cause a disproportionately high mortality close to adult trees could explain the high species diversity (on a per area basis) in tropical forests. Lowland tropical rain forests in western Amazonia and Borneo (Gentry 1988a) can have more than 300 tree species with >10 cm dbh in a single hectare, and the evenness of tree species abundances is also high in such forests (Gentry 1988a; Wright, Chap. 2, this Vol.). Clark and Clark (1984) found density-dependent mortality of *Dipteryx panamensis* seedlings in lowland Costa Rica that was consistent with the Jansen-Connell escape hypothesis, and could not be explained by self-thinning or inhibition by conspecific adults. Furthermore, Clark and Clark (1984) found that a preponderance of studies on woody plants had results that were consistent with density- or distance-dependent mortality in both wet and dry tropical forests, though the mechanism could not be identified in most cases.

A few studies have shown the importance of fungal pathogens in determining tropical tree dispersion patterns through their effects on seed germination, and seedling and sapling survival. Burdon and Chilvers (1982) found evidence of density dependence in 39 of the 69 studies of fungal diseases they reviewed (Table 5.1). Five of the nine seed studies reviewed by Clark and Clark (1984) were consistent with density- or distance-dependent mortality from pathogens and seed predators. Augspurger (1983) found that higher proportions of *Platypodium elegans* seedlings in Panama survive damping-off caused by nonspecific fungal pathogens at distances further away from the parent plants. Seedlings of *Theobroma* and *Herrania* spp. in the Amazon and Orinoco basins of South America are sometimes clumped together in forest gaps because they grow more rapidly in full sun, but because unshaded saplings with rapidly growing meristems are killed by the Witches' Broom fungus, only scattered trees in the forest understory are usually recruited into the adult population (H. C. Evans, in Lodge and Cantrell 1995). Gilbert et al. (1994) found that the incidence of a fungal canker disease on *Ocotea whitei* on Barro Colorado Island, Panama, was host-density dependent, resulting in a net spatial shift of the juvenile population away from conspecific adults. These and other diseases may affect nutrient cycling in tropical forests by reducing clumping of their hosts.

5.5 Microbial Contributions to Global Biogeochemistry

Microorganisms have had a tremendous influence on the availability of nutrients in soil through their weathering of rocks, and on the composition of the Earth's atmosphere, especially the increases in greenhouse gases such as carbon dioxide, nitrous oxide, and methane. Nitrous oxide contributes to destruction of the Earth's ozone layer, in addition to its role as a greenhouse gas. Although the roles of microorganisms in production of greenhouse gases are global and not specific to tropical forests, it is appropriate to draw attention to fluxes that have tropical foci.

The consequences of a reduction in microbial diversity on the fluxes of greenhouse gases is unknown. There are apparently few species of methane oxidizing and autotrophic denitrifying bacteria (Payne 1973; Frobisher et al. 1974), so the processes they mediate could be susceptible to perturbations that reduce diversity. The mechanisms and roles of specific microorganisms in the decomposition of soil organic matter is largely unknown (Jorgensen et al. 1990).

5.5.1 Atmospheric CO_2

Microbial decomposition of soil organic matter (SOM) in the tropics can be a significant source of global CO_2. A flux of 1.6 Pg yr^{-1} was recorded from low latitude forests for 1987-90 (Dixon et al. 1994). Organic matter contents in temperate and tropical soils are comparable (Sanchez et al. 1982), and tropical forests represent 27% of the global soil carbon pools (Dixon et al. 1994). The organic-matter contents of tropical soils that have been converted to agriculture generally decline within a few years to 30 to 64% of the initial content under natural vegetation (Young 1976); declines in SOM did not occur with conversion of forest to pasture in Brazil (Bonde et al. 1992). Thus conversion of large tracts of tropical forest to agriculture can cause a short-term increase in the global flux of CO_2. Bonde et al. (1992) studied the decomposition of SOM following the conversion of Brazilian Amazonian forest to sugarcane. During 12 years of cultivation, 83 and 93% of the carbon derived from forest that was associated with silt and sand fractions had decomposed, whereas only 40% that was associated with clay had decomposed. These data suggest that silty and sandy tropical soils are especially susceptible to losses of soil organic carbon following deforestation.

5.5.2 Methane

The amount of methane released globally into the atmosphere from natural wetlands on an annual basis is 110 Tg (Bartlett and Harriss 1993; Mathews and Fung 1987), primarily from the activity of methanogenic bacteria (Fig. 5.1A; Table 5.1). Sixty percent of the total global natural methane emission is from tropical wetlands (Bartlett and Harriss 1993). In the tropics, 28-31 Tg (25-28% of the total natural) is emitted from forested wetlands, while about 36 Tg (33% of the total natural) is emitted from nonforested wetlands (Bartlett and Harriss 1993). The total global flux of methane from all sources including anthropogenic and geochemical is about 400-500 Tg per year. Thus, methane emissions from all types of tropical wetlands constitute 13-17%, and forested tropical wetlands contribute 6-8%, of the total global methane emissions. Although the intestinal microbiota in termites contribute to methane emissions from tropical forests outside wetlands, the soils in these well-drained sites are often net sinks for methane, and may thus largely offset methane emissions from termites (Seiler et al. 1984). Conversion of upland forest to agriculture, however, could disturb microbial consumers of methane and reduce the effectiveness of these methane sinks.

5.5.3 Nitrous Oxide

The total flux of nitrogen as nitrous oxide into the atmosphere is 14 Tg on an annual basis (Prinn et al. 1990). Moist and wet tropical forests emit about 3 Tg of nitrous oxide-N annually (Figure 5.1B; Matson and Vitousek 1990). Microbial denitrification is primarily responsible for the terrestrial losses of nitrogen through nitrous oxide emissions, but nitrification also contributes an unknown portion. Early succession after clearing of tropical wet forests often results in a pulse of denitrification, followed by a decline during mid-succession because of competition for nitrate (Robertson and Tiedje 1988).

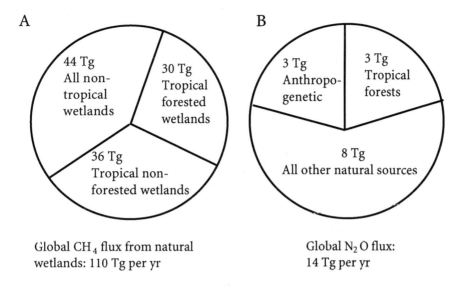

Fig. 5.1. A Methane released globally from natural wetlands (Bartlett et al. 1993; Mathews and Fung 1987). **B** Nitrous oxides released globally from all sources; 1 Tg = 10^{12} g. (Matson and Vitousek 1990)

5.5.4 Rock Weathering

Often the most important source of certain mineral nutrients that are necessary for plant growth, such as potassium, calcium, and magnesium, is from weathering of rocks. The various processes involved in microbial transformation of rocks and soil formation have been studied mainly in temperate zones, but there is no reason to suspect that rocks in the tropics are affected differently (Table 5.1). Many lichenized fungi grow on (epilithic) and in (endolithic) rock and degrade the substrate both by mechanical action of hyphae that penetrate the first few millimeters of the surface, and by chemical action (especially the conversion of silicates to oxalates; Jones and Wilson 1985). Algae develop on porous stone that is warm and damp. Griffin et al. (1991) suggested that algae degrade stone by both mechanical and chemical means. Algal communities on stone are often embedded in surface slimy mats together with heterotrophic bacteria which breakdown trapped organic matter, and which undergo considerable volume changes through repeated wetting/drying cycles. This has the effect of loosening the stone particles (Ortega-Calvo et al. 1991). Among fungi, Mucorales, ascomycetes, and hyphomycetes are the main groups that decay rock by mechanical and chemical means. Fungi in association with bacteria are widespread on deteriorating stone, and their hyphae are able to penetrate deeply (Lepidi and Schippa 1973).

Three different groups of bacteria have been associated with rock decay; autotrophic sulfur-oxidizing, nitrifying, and heterotrophic (Hueck van der Plas 1968; Table 5.1). Phototrophic cyanobacteria have also been implicated (Ortega-Calvo et al. 1991). Sulfur bacteria are chemolithotrophs; their energy is generated from the metabolism of reduced inorganic sulfur compounds (sulfide, thiosulfate, and elemental sulfur) to sulfate which, as sulfuric acid, can cause severe damage to mineral materials. Nitrifying bacteria, also chemolithotrophs, oxidize inorganic nitrogen compounds for energy and generate acidic end products. Heterotrophs are thought to play an important role in the weathering of rocks and minerals during soil formation and its subsequent fertility (Berthelin 1983).

5.6 Nutrient Cycling

5.6.1 Litter Decomposition and Soil Fertility

The rate of decomposition of organic matter influences the rate at which nutrients, especially nitrogen and phosphorus, are mineralized and potentially become available for plant growth. The turnover time for fine litterfall is usually shorter in lowland tropical as opposed to temperate forests, but there is considerable overlap in rates of decomposition between the slowest decomposing tropical leaf species and the fastest decomposing temperate leaf species (Burgess 1960; Olsen 1963; Anderson and Swift 1983). Differences in rates of decomposition are thought to be related to the nature of the decomposer community, the characteristics of the organic matter (i.e., resource quality), and the physical and chemical environment (Swift et al. 1979; Anderson and Swift 1983). Although these factors may interact, and their relative importance can vary between wood, leaves, and fruits, they are thought to have the following general hierarchial structure in the way they influence each other: macroclimate > microclimate > resource quality > organisms.

Anderson and Swift (1983) surmised that variations in the composition of the "microflora" (i.e., fungi and bacteria) were generally not important in determining litter decay rates because there are many species of fungi and bacteria that decompose litter and that various combinations of microbes were likely to have the same functional capacity (Table 5.1). For example, Bills and Polishook (1994) cultured 81-125 species of microfungi from each of their small samples (about 0.25 m^2) of rain forest leaf litter in Costa Rica. Although 200 species of microfungi were cultured in that study, the fungal community was greatly undersampled (Bills and Polishook 1994; Lodge 1985). Cornejo et al. (1994) cultured 500 species of microfungi from the leaf litter of six tree species in seasonally dry forest in Panama.

Anderson and Swift (1983) may be correct in assuming that the microbiota is relatively unimportant in determining the rates of litter decomposition, but a few temperate studies suggest otherwise, and relevant data from the tropics are scarce. As discussed earlier, preferences among invertebrates for feeding on certain decomposer fungi have been found to influence the rates of decomposition and nitrogen mineralization from leaf litter in temperate forests (Newell 1984) and streams (Arsuffi and Suberkropp 1984). Disturbances in tropical forests have been found to affect fungal communities and fungivorous invertebrates (Lodge and Cantrell 1995). Although leaf litter in Panama had a high diversity and a low dominance of microfungi (Cornejo et al. 1994), microfungal species had preferences for leaves of certain tree species and their responses to increased moisture were individualistic. The loss of microfungi that are host-specific would probably not affect the rates of litter decomposition, but some of these fungi are endophytic and colonize leaves before they fall (Bills and Polishook 1994; Cornejo et al. 1994), and they may be important in immobilizing nutrients and slowing leaching losses from freshly fallen litter.

Certain macrobasidiomycete species are potentially more important in determining the rates of decomposition and nutrient mineralization, and they may be more vulnerable to disturbances than microfungi. Basidiomycetes can be very abundant in the litter layer (Hedger 1985; Lodge and Asbury 1988) and species that have superficial or interstitial mycelium are especially sensitive to moisture stress. Canopy openings (Hedger 1985; Lodge and Cantrell 1995) and salt stress from fertilization (D.J. Lodge, S. Cantrell, and O.O. Molina-Gomez, unpubl. data) have caused changes in litter basidiomycete community composition and reductions in species richness. The loss of basidiomycete fungi with superficial or interstitial mycelia from the litter layer following disruption of the canopy by Hurricane Hugo is thought to have contributed to the subsequent decline in fungivorous snails in Puerto Rico (Willig and Camilo 1991). It is not known if nutrient cycling was affected by this loss, but a dominant species in this guild, *Collybia johnstonii*, was severely affected (Lodge and Cantrell 1995) and was known to be important in translocating nutrients in the litter layer (Lodge 1993) and in slowing losses of nutrients from the ecosystem by reducing erosion (Lodge and Asbury 1988). Basidiomycete fungi that translocate limiting nutrients from decomposed to fresh resources are thought to accelerate the decomposition of low quality substrates such as wood (Boddy 1993) and some tropical leaf litter (Lodge 1993; Lodge et al. 1994). In addition, decomposer fungi that translocate nutrients in litter can influence the fate of nutrients in the ecosystem by accentuating pulsed nutrient releases (Lodge et al. 1994).

Yang and Insam (1991) observed that about 75% of the soil microbial biomass in Hainan Island, China, was bacterial, but fungi dominate soil microbial biomass in other tropical sites such as in Puerto Rico. Various studies indicate the importance of microbial biomass in nutrient retention and regulation of nutrient availability in tropical forest ecosystems (review by Lodge 1993), but there are no studies relating soil microbial diversity to these ecosystem processes. Drying/rewetting cycles can accelerate the replenishment of the available soil nitrogen pool from microbial, recalcitrant, or physically protected nitrogen pools (Cabrera 1993). Fluctuations in moisture cause periodic crashes in populations of the soil microbial biomass, inducing pulses of nutrient release (Raghubanshi et al. 1990; Srivastava 1992; Davidson et al. 1993; Lodge 1993). These cycles in soil nutrient availability and moisture may increase the uptake of limiting nutrients by plants when pulses are synchronized with plant uptake, whereas asynchrony can lead to large losses of nutrients from the ecosystem or the biologically available pools (Lodge 1993; Lodge et al. 1994). Global climate change may influence the rates of nutrient mineralization and the fates of those nutrients in tropical forests in part because it is expected to alter the frequency and intensity of rainfall at tropical lattitudes.

5.6.2 Symbiotic Nitrogen Fixation Associated with Plant Roots

Ninety million tons of nitrogen are made available to the biota globally on an annual basis by nitrogen-fixing microorganisms, primarily those associated with plant roots (Curtis 1975; Table 5.1). Nitrogen-fixing *Rhizobium* bacteria in nodules on the roots of leguminous forest trees are an important means of maintaining such plants in nutrient-poor tropical soils (Sprent and Sprent 1990). Edmisten (1970) estimated that root nodules contributed 0.27 kg N ha^{-1} yr^{-1} to native subtropical wet forest in Puerto Rico. Symbiotic nitrogen fixation in other tropical forests that are dominated by leguminous trees is probably much greater, such as in some forests in parts of Africa and South America. A few tropical trees such as *Casuarina* in Australia (Torrey 1982) and *Alnus* in the Central and South American Andes form mutualistic root associations with actinomycete bacteria, known as actinorrhizae. The importance of actinorrhizae has been demonstrated in temperate and subtropical areas (Rose and Trappe 1980; Rose and Youngberg 1981; Esteban et al. 1987; Schmitz et al. 1990; Vitousek 1990), but there are few such studies in tropical forests. A notable exception is the study by Vitousek and Walker (1989) that documented alterations in forest ecosystem processes which resulted from increased nitrogen availability following the introduction to Hawaii of the invasive actinorrhizal species *Myrica faya*.

A loss of diversity of *Rhizobium* species and strains could severely limit the input of biologically available nitrogen, especially in tropical forests that have many leguminous tree species in the Mimosaceae and Papilionaceae (the majority of Caesalpinaceae do not form nodules). Inappropriate combinations of host species or genotypes with *Rhizobium* species or serotypes often result in lack of inoculation or an inability of nodules that are formed to fix nitrogen. An insufficiency of appropriate N-fixing bacteria is likely to occur during recolonization of a disturbed area because the bacteria are not disseminated with the seeds of their host. In contrast, the addition of a single nitrogen-fixing actinomycete species and its host to an area where N-fixing symbioses were absent in the native flora has caused a massive disruption of the forest and its ecosystem processes in Hawaii.

5.6.3 Effects of Microbial Epiphylls and Epiphytes on Nutrient Fluxes

Dense mats or mosaics on the leaves of evergreen forest trees, consisting of sooty molds, melioline and asterine fungi, filamentous algae, and folii-colous lichenized fungi are especially abundant in cloud forests and forest understories, and can affect cloud capture, the quantity and nutrient content in throughfall, and net photosynthesis. Together with other foliicolous and caulicolous epiphytes, epiphytic lichens and microbial growths enhance capture of cloud water and slow the flow of water through the canopy (Weaver 1972; Knops et al. 1992). Data on the effects of epiphytic lichens on the quantity and quality of throughfall are not yet available from the tropics. The removal of epiphytic lichens from an oak woodland in coastal California, however, decreased canopy interception of rainfall by half (Knops et al. 1992). In addition, lichens were found to augment the fluxes of NO_3-, organic-N, Ca^{2+}, Mg^{2+}, Na^+, and Cl^- in throughfall (Knops et al. 1992).

Macrolichens are particularly conspicuous in montane tropical forests, and include species in groups known to be indicators of forest disturbance in temperate forests (Rose 1992). Field observations in Australia, Mexico, Trinidad and Tobago (D. L. Hawksworth, unpubl. data), and Thailand (Wolseley and Aguirre-Hudson 1991) indicate that the same phenomenon occurs in tropical forests, although different species can be involved.

Although understory leaves with a heavy epiphyll cover probably have negative carbon gains (they represent a net energy loss to the plant), these leaves may be retained by the plant because they absorb nitrogen released by the nitrogen-fixing epiphylls (Bentley and Carpenter 1984; Bentley 1987). Some of the nitrogen released from epiphylls upon rewetting following a drying phase is absorbed by the leaves supporting them (Edmisten and Kline 1968; Bently and Carpenter 1980; Edmisten 1970), and some contribute significantly to nitrogen flux in throughfall and stem-flow (Edmisten 1970; Coxson 1991). A diversity of epiphytic bacteria, yeasts, cyanobacteria, and cyanobacteria-containing lichens is involved in fixation of atmospheric nitrogen (Table 5.1; Whitton 1992), and their nutrient contributions may be significant in some tropical forests. In a study of a Colombian rain forest canopy, cyanobacterial lichens contributed 1 to 8 kg ha^{-1}

of nitrogen to the forest ecosystem each year (Forman 1975). Sheriden (1991) estimated that epicaulous cyanobacteria in mangrove forests of Guadeloupe, French West Indies, fixed 4.2 kg N ha^{-1} yr^{-1}. In a wet forest in Puerto Rico, Edmisten (1970) estimated that nitrogen fixation by epiphylls contributed 0.6 kg N ha^{-1} yr^{-1}, and were the most important single source of nitrogen in that forest. Fritz-Sheridan and Portecop (1987) found very high rates of nitrogen fixation in cloud forest epiphylls on the volcano La Soufriere on Guadeloupe, but did not estimate the resulting total nitrogen input to the ecosystem.

In lowland tropical rain forests, epiphytic nitrogen-fixing cyanobacteria and cyanobacterial lichens are most abundant in the humid understory (Akinsoji 1991), which suggests that they are sensitive to desiccation or high light. In Costa Rica, elegant studies by Lücking (1992) showed that forest disturbance altered the assemblages of different leaf-dwelling (foliicolous) lichens. Lichens are damaged by sulfur dioxide air pollution in the tropics, as they are in temperate regions. A reduction or loss of the nitrogen-fixing epiphyte guild could significantly reduce total nitrogen inputs in some forests, but data on the relative importance of this source are needed from a greater diversity of tropical forests to determine if the threat is widespread.

5.6.4 Mycorrhizae and Nutrient Uptake

Mycorrhizal fungi are symbiotic with plant roots, and improve plant growth and survival in nutrient-poor soils by facilitating the uptake of water and limiting nutrients, especially phosphorus and nitrogen (Bowen 1980; Janos 1980, 1983; Harley and Smith 1983; Table 5.1). Mycorrhizae can also increase the resistance of plants to root pathogens and improve their tolerance of toxic metals (Harley and Smith 1983). Most tropical plants, including forest trees, form vesicular-arbuscular mycorrhizae (VAM) with zygomycetous fungi (Janos 1983). Some tropical trees (e.g., Diptero-carpaceae and some species of Leguminoseae (Caesalpinaceae and Papil-ionaceae), Nyctaginaceae, Myrtaceae, Euphorbiaceae, Polygonaceae, and Proteaceae (Redhead 1980; Janos 1983; Tacon et al. 1989; Thoen 1993), especially those growing in very nutrient-poor soils or areas with seasonal drought, form ectomycorrhizae with mostly basidiomycete fungi. Another type of sheathing mycorrhiza found on some tropical trees is also formed with basidiomycete fungi, but the ecological significance of this association has not been studied (Ashford and Allaway 1982, 1985; Lodge 1987a). Mycorrhizal fungi are crucial for seedling establishment in orchids, and the net flow of carbohydrates from the fungus to the orchid often continues after the orchid is photosynthetic (Harley and Smith 1983). Many plants that lack chlorophyll, such as species of *Coralorhiza, Burmania, and*

Wullschlegiella, obtain all of their nourishment from a mycorrhizal fungus (Harley and Smith 1983; Perry 1990).

Mycorrhizal fungi are a possible keystone guild with regard to nutrient cycling and primary productivity in topical forests. There are almost no studies of the relationship between the diversity of mycorrhizal fungi per se and ecosystem processes, but a number of studies suggest that mycorrhizal fungi may be vulnerable to disturbance, and a reduction in their diversity might reduce the rate of forest recovery. In areas where the upper layers of soil, together with its complement of mycorrhizal fungi, has been eroded away, the capacity of the soil to establish and maintain healthy plant populations can be severely limited (Powell 1980). In Cameroon, West Africa, complete clearance of forest sites before replanting with native species resulted in a 65% reduction in VAM spores; spore populations were less damaged by partial site clearance. The loss of mycorrhizal fungi slows down the reintroduction/regeneration of trees that are obligately mycotrophic (Janos 1983). Spread of VA-mycorrhizal fungi in tropical forests occurs primarily from root-to-root by hyphal growth over distances of millimeters to centimeters because spore production and survival is low (Janos 1983). Although some ectomycorrhizal fungi may colonize more readily than VAM fungi, reintroduced seedlings of dipterocarps in Malaysia often remained uninfected for 6 months or more when not in contact with living mycorrhizal roots in early (within 20 days of germination) stages of growth (Alexander et al. 1992).

In Mexico, over-collection due to increased popularity of "wild" edible fungi is reported to cause a significant decrease in sporophore production in forest areas (Villarreal and Perez-Moreno 1989). Similar declines in Europe were experimentally demonstrated to be caused by trampling rather than removal of the fungal fruiting bodies. The ability of seedlings to colonize beyond the forest edge might become limited by a reduction in sporophores of ectomycorrhizal fungi.

In addition to reducing the overall abundance of mycorrhizal fungi, disturbances have been shown to change species composition in VAM communities. For example, Wilson and Mason (1993) studied a chronosequence of forest plantations in the Ivory Coast and found that the balance of VAM fungal species was still perturbed 23 years following clearance and replanting. The effects of such changes in VAM communities have not been studied in tropical forests, but there are recent suggestions in the temperate literature that a host may not benefit from mycorrhizae formed with an inappropriate fungus. For example, wheat infected by the dominant mycorrhizal fungus from a virgin prairie soil elicited more negative leaf osmotic potentials, which allowed positive turgor pressure to be maintained, but when wheat was infected by the dominant mycorrhizal fungus from a tilled soil there was no similar response to drought stress, resulting in reduced vegetative growth (Miller 1989). The global diversity of VAM fungi is low, and despite their ability to colonize a wide diversity of trees in tropical forests (Malloch et al. 1980), the data of Miller (1989) suggest that a loss of diversity of VAM fungi could have a negative effect on primary productivity.

Mycorrhizal fungi may be susceptible to fungicides which are often applied to tropical forests where coffee, cacao, and other intensively managed crops are planted in the understory of native forests. Most studies on the effects of fungicides on ectomycorrhizal fungi have been either in vitro laboratory experiments, to assess their use in fungal identification (Hutchison 1990), or on conifers in temperate and subtropical nurseries (Tacon et al. 1986; Kelley 1987). However, Unestam et al. (1990) found that the results of in vitro testing of the effects of five fungicides on four ecto-mycorrhizal fungi did not closely resemble those from in vivo studies on pot-grown seedlings. We did not find any studies on tropical forest systems.

Many basic questions about the roles of ectomycorrhizal fungal species in tropical forests cannot be answered as yet (Thoen 1993). Are ectomy-corrhizae important for seedling survival? Are different ectomycorrhizal fungi required as the host ages? Sequences of ectomycorrhizal fungi with tree age occur in both temperate and tropical forests (Last et al. 1992). What is the influence of forest fires on ectomycorrhizal communities? How are ectomycorrhizal fungi dispersed? Early stage ectomycorrhizal fungi are able to colonize seedlings via spores whereas late-tage fungi are not (Deacon and Fleming 1992; Last et al. 1992), suggesting that spores of late-stage fungi may function in gene flow between populations rather than in colonization (Egger 1995). If late stage fungi are required, then ectomy-corrhizal trees may not thrive beyond a certain age in deforested areas that have been reforested. Are ectomycorrhizal fungi able to survive distur-bance, and if so, for how long? Do ectomycorrhizal fungi decompose litter and transer the nutrients to their host trees? Research to answer these questions will be needed before we can speculate on the potential impacts of a reduction in ectomycorrhizal fungal diversity on ecosystem processes.

5.7 Plant Endophytes

Mutualisms between fungi living inside plants (endophytes) and their hosts are called mycophyllas, and are ubiquitous. It has been estimated by M. Dreyfuss that each tropical forest tree has three to four characteristic endo-phytic fungi (Hawksworth 1991). In a study by Laessøe and Lodge (1994) in Puerto Rico, nine endophytic xylariaceous fungi were isolated from healthy petioles of *Schefflera morototoni*. Similarly, Rodrigues et al. (1993) found 15 different xylariaceous fungi growing in healthy green leaves of the palm *Euterpe oleracea* in Amazonas, Brazil. In both studies, however, the identi-fiable *Xylaria* species were ones with broad host ranges. Physiological races adapted to specific hosts may have evolved in these widespread endophytic fungi, but this aspect has not been investigated. The role of mycophyllas is unclear, but at least in well-studied cases from the temperate zone, the endophytic fungus produces secondary metabolites that deter foraging organisms and pests (Clay 1988). A tropical endophytic *Xylaria* species was studied in vitro and found to be an aggressive antagonist of the fungus that causes Witches' Broom Disease in *Theobroma* and *Herrania* species (Bravo-Velasquez and Hedger 1988). We cannot speculate on the vulnerability of endophytic fungi or what effect a reduction in their diversity would have on ecosystems without studies on their significance to tropical trees, their host ranges, and their means of dispersal (Lodge and Cantrell 1995).

5.8 Threats to the Microbiota and the Processes They Mediate

Threats to microbial diversity in tropical forests include forest fragmentation, loss of hosts caused by logging and other human activities, air pollutants, fungicides, disturbances that alter microclimates and exposure to sunlight, global climatic changes that alter the frequency of rainfall and extreme events such as droughts, and the direct or indirect effects of over-exploitation of edible fungi. We can only speculate on what impact a loss of microbial species richness or the loss of an entire functional group would have on tropical forest ecosystems based on the importance of the processes they mediate and their sensitivities to disturbance.

5.8.1 Effects of Forest Fragmentation on Plant Symbioses

Problems of co-dispersal in a fragmented landscape can alter nutrient cycling by threatening plants that depend on root symbioses with nitrogen-fixing bacteria and VAM fungi for nutrient transfers across the plant-atmosphere and soil-plant interfaces. Forest fragmentation may contribute to extinctions of symbiotic microbes that are host-specific and have poor dispersal abilities, such as some late stage ectomycorrhizal fungi and possibly some plant endophytic fungi. Some evidence suggests that symbiotic VA-mycorrhizal fungi and nitrogen-fixing bacteria are often adapted to a narrow range of environmental conditions, but research in tropical forests will be needed to determine conclusively if disturbances such as deforestation and overgrazing alter the soil environment so drastically that mycorrhizal fungi from the neighboring forests are no longer capable of colonizing and surviving. At least in the case of mycorrhizal fungi, "redundancy" imparts some resilience to perturbations to the ecosystem, as discussed by Perry et al. (1989). The limited abilities of VAM fungi to disperse on ecological time scales may have evolutionarily favored their ability to associate with many plant species (Janos 1983; Miller 1989); similarly, nitrogen-fixing bacteria can often form nodules with more than one host plant genotype. Only certain combinations of plant/microbe species or genotypes may benefit the plant, however, so the loss of a particularly effective and efficient microbial symbiont because of environmental changes or lack of codispersal could reduce productivity in trees that are dependent on them. Research is needed to determine if reductions in the abundance of mycorrhizal fungi and the shifts in VA-mycorrhizal fungal species composition that have been observed following disturbance in tropical forests, cause a reduction in host plant productivity or a shift in plant community composition.

Basic research is needed to determine if mycophyllas (mutualism between endophytic fungi and their host plants) have a keystone effect in tropical forest trees. Evidence from the temperate zone suggests that some fungal endophytes produce potent compounds that protect their host from herbivory, but little or nothing is known about the role of these fungi in tropical forests, where they are reputedly abundant. We also do not know the host ranges of these endophytic fungi, or if the variation seen among isolates of what appears to be the same morphospecies, is significant and correlates with host specificity (Lodge and Cantrell 1995). If the host range of these fungi is truly broad, then, as in the case of nonspecific mycorrhizal fungi, they are more likely to make the ecosystem more resilient to disturbance than if they are adapted to particular hosts. On the other hand, if their host ranges are narrow, then a "reduction" in their host's density may present a special problem in terms of fungal colonization. If tropical mycophyllas play a crucial role in plant defence against herbivory, as they do in temperate systems, and if they are host-restricted, then reduction of host density could lead to loss of the endophytes and subsequent extinction of their host plants (Lodge and Cantrell 1995).

5.8.2 Effects of Forest Fragmentation on Cord-Forming Fungi

Colonies of cord-forming decomposer basidiomycetes can be extensive, and could therefore be affected by forest fragmentation, roads, and trails. Some cord-forming fungi that kill trees can extend over many hectares (Kile 1983; Smith et al. 1992), are often favored by disturbances such as droughts, selective harvesting, and agroforestry (Edgar et al. 1976; Fagg et al. 1986; Castello et al. 1995), and are themselves a major cause of forest disturbance(Castello et al. 1995). Nonpathogenic cord-forming fungi are another potentially keystone guild because they can accelerate the rate of wood decomposition by translocating limiting nutrients over large distances from previous resource bases (Boddy 1993). Typically, wood-decomposer fungal colonies of this type extend up to 10-40 m in temperate and tropical forests (Thompson and Boddy 1983; Dowson et al. 1989a, b; R.G. Bolton and L. Boddy, in Boddy 1993). Immobilization of nutrients resulting from the activities of wood decomposers in fresh debris can buffer a forest ecosystem from nutrient losses following disturbances, such as logging (Matson and Vitousek 1981) and hurricanes (Zimmerman et al. 1995), so the removal of wood for fuel, lumber, and other forest products can greatly accelerate nutrient losses by reducing fungal activity. However, cord-forming fungi may reduce primary productivity (Zimmerman et al. 1995) by directly or indirectly limiting the availability of mineral nutrients to trees (Wells and Boddy 1990; Boddy 1993; Lodge 1993; Lodge et al. 1994).

5.8.3 Effects of Acid Precipitation on Ectomycorrhizae

Acid precipitation is believed to be the cause of mass extinctions of European ectomycorrhizal fungi (Arnolds 1991; Jaenike 1991; Fellner 1993), so tropical forests may also be at risk from both local and long-distance air pollution sources. However, cause and effect have not yet been incontrovertibly demonstrated. Some of the losses of ectomycorrhizal fungi in Europe may relate to succession of fungal communities with aging of their hosts or development of forest soils rather than acid precipitation. Research is needed to determine if tropical ectomycorrhizal trees require different species of fungi at different ages or in different soils.

5.8.4 Effects of Air Pollution and Climate Change on Epiphyte Nitrogen Fixation

Nitrogen-fixing epiphytes are a group of potentially sensitive microorganisms that may play a keystone role in nutrient cycling in tropical forests. The annual flux of nitrogen fixed by arboreal free-living and lichenized cyanobacteria has been studied at only a few tropical forest sites, but the data suggest that the amounts of nitrogen fixed can be substantial relative to other sources. Because of their direct exposure to the atmosphere, epiphytes are particularly sensitive to changes in humidity, the distribution and amount of precipitation, air pollutants, and increased UV radiation (Galloway 1991; Wolseley and Aguirre-Hudson 1991). Lichens are frequently used as bioindicatiors because of their susceptibility to air pollution, especially sulfur dioxide (Nash and Wirth 1988; Richardson 1987, 1992). There is a sufficient body of pertinent work to demonstrate that effects similar to those documented in temperate regions can be expected in the tropics (e.g., Thrower 1980; Hawksworth and Weng 1990; Galloway 1991), but we know of no studies in natural tropical forests. Loss of these organisms would affect other organisms dependent on them for atmospheric nitrogen-fixation, food, camouflage, or shelter. However, too little is known about the general importance of cyanobacteria in the overall nitrogen budgets of tropical forests, and even less about their susceptibility to stress to generalize.

5.8.5 Are Decomposers Redundant in a Heterogeneous Environment?

In general, a variety of microbial decomposer species with apparently similar biologies can be found in a single ecosystem. It is therefore tempting to speculate that the loss of some of these would have little impact on the functioning of an ecosystem as a whole, provided that a large enough population of at least one species remained to carry out that process for the whole ecosystem. However, we rarely know if the apparently similar functions performed are truly identical. For example, although a range of macrofungi can be involved in the decomposition of a single log, the hypothesis that each might have enzymes capable of breaking different ligno-cellulose bonds cannot be discounted on the basis of the data currently available. Similarly, different macromycetes decompose leaf litter at different depths and levels of breakdown (Hedger 1985). In addition, host specificity among some groups of fungal decomposers contributes to their high diversity in tropical forest litter (Holler and Cowley 1970; Cornejo et al. 1994; Laessøe and Lodge 1994) and may indicate an ability to tolerate or decompose particular plant-defense compounds (Bharat et al. 1988). Finally, natural and anthropogenic disturbances in tropical forests create patches that differ in temperature, moisture, and mineral nutrient regimes, and consequently have different communities of decomposer fungi (Hedger 1985; Castillo-Cabello et al. 1994; Lodge and Cantrell 1995). In the Ivory Coast, soil fungal diversity and ecosystem processes recovered quickly from forest disturbance caused by shifting cultivation (Maggi et al. 1990). The most important factor controlling fungal diversity and community composition in that study was seasonality of rainfall, but differences among soil types became important in unusually dry years (Maggi et al. 1990). Thus, diversity among decomposer fungi may impart some resiliency to spatial and temporal heterogeneity in environments resulting from natural and anthropogenic disturbances.

5.9 Conclusions

The minimal attention accorded to microorganisms in tropical forests to date is not proportional to the multiplicity of crucial ecosytem-level processes they mediate. For this reason, it is impossible in most cases to know the extent of functional redundancy among microorganisms, and thus to predict with certainty the impact of a loss of microbial diversity on overall ecosystem processes, or the resisiliency of those processes following disturbance. There are sufficient indications, however, that some functional groups of microorganisms that are sensitive to stress play fundamental roles in ecosystem processes, which justify their further study and the monitoring of the key processes they mediate within conservation programs. In Table 5.1 we have identified these key processes, the mechanisms, and the groups of microorganisms involved.

Keystone species are likely to occur among symbionts that mediate key processes because host specificity may preclude substitutions among species of the same functional group. Examples of potentially keystone mutualistic symbionts include mycorrhizal fungi, nodule-forming nitrogen fixers, arthropod gut microorganisms, and possibly endophytic fungi. Host specificity is also common among pathogens of herbivores, and these parasites may play a keystone role in limiting secondary production. Host-specific microbial symbionts that are at greatest risk from anthropogenic disturbances are those that lack codispersal with their host and are also slow to disperse on ecological time scales, such as late stage ectomycorrhizal and VAM fungi, nodulating bacteria, and actinomycetes.

Irrespective of their specificity for particular hosts, certain keystone fungi pathogenic to trees can directly influence primary production and the structure of tropical forests by causing gaps. The most severe examples of this type were aggravated by anthropogenic disturbances or initiated by human introductions. Tropical forests with a low diversity of tree species or genera (e.g., palm and eucalypt forests) are at greater risk than those with a higher diversity of tree taxa. Plant pathogens can have more subtle, indirect effects on ecosystem-level processes, such as decomposition and nutrient cycling, by changing the distribution and dispersion of tree species, and increasing the prevalence of secondary plant defense compounds.

Some ecosytem-level processes are vulnerable to disturbances in tropical forests because the functional group of microorganisms involved or the processes they mediate are confined to the same narrow environment or are sensitive to the same stresses. For example, although a moderate number of microbial species is involved in production and consumption of various trace gases (e.g., denitrifiers, methanogens, and methane consumers), these functional groups or the processes they mediate are sensitive to environmental changes and substrate availability. Arboreal autotrophic and heterotrophic nitrogen fixers are especially vulnerable to air pollutants because they are exposed directly to the atmosphere and they are adapted to scavenging nutrients from the air and rain. This group may play a keystone function in some tropical forests, where they are the greatest source of mineral nitrogen.

The processes that might be least vulnerable to losses of microbial species are those in which considerable numbers of microorganisms are involved, including decomposition and nutrient mineralization, nitrification, and the maintenance of soil structure. Although we may classify microorganisms as belonging to the same functional group (Table 5.1), many members of a group cannot replace another species, and they are therefore not truly redundant, because they occupy different niches as defined by their host preferences or tolerance of drought, rewetting, anoxia, high temperatures, low nutrients, high salts, or secondary plant compounds and humic acids. Thus, some of the diversity within large functional groups, such as decomposers, is required for maintenance of ecosystem processes in the heterogeneous environments of tropical forests, and their diversity becomes more important when those ecosystems are subjected to stress (Lodge and Cantrell 1995).

Most attention has been given to what effect the loss of microbial diversity would have on a given ecosystem process, but the introduction of a novel functional group to an insular ecosystem can also be devastating. The introduction of a plant with its nitrogen-fixing actinomycete root symbiont to Hawaii, where no such associations had previously existed, has changed forest community composition, forest structure, and nitrogen cycling (Vitousek and Walker 1989; Vitousek 1990).

The functional processes that merit the most attention in the future are those uniquely or predominantly mediated by microorganisms and on which other organisms ultimately depend. Use of morphological, cultural, chemical, and molecular traits may be useful in assessing microbial diversity because naming all of the microorganisms within functional groups may be impractical, faced with so many undescribed species and limited human and financial resources (Hawksworth and Ritchie 1993). Given the situation, it is important for the few microbiologists working in the tropics to identify and prioritize research needs. Although a discussion on the selection of these priorities falls outside the scope of this chapter, and is considered elsewhere (Hawksworth and Ritchie 1993), we encourage bacteriologists and mycologists to collaborate with tropical ecologists so that more comprehensive data sets can be accumulated for particular sites, including measurable parameters of functional attributes such as rates of nitrification, denitrification and decomposition (Hawksworth and Colwell 1992). Only in this way will a more quantitative and objective assessment of the role of microbial diversity in ecosystem function eventually be realized.

Acknowledgments. We thank the ecologists participating in the Oaxtapec workshop in December 1993 for their stimulating suggestions, and especially Drs. G. H. Orians and R. Dirzo for constructive criticism and suggestions on the text. We are indebted also to Mrs. L. Wheater, who searched the CAB ABSTRACTS and BIOSIS databases to locate pertinent publications, and to Mrs. G. Reyes of the Library of the International Institute of Tropical Forestry, USDA Forest Service in Puerto Rico for obtaining reprints.

References

Akinsoji A (1991) Studies on epiphytic flora of a tropical rain forest in southwestern Nigeria: II: Bark microflora. Vegetatio 92:181-185

Alexander I, Ahmad N, See LS (1992) The rôle of mycorrhizas in the regeneration of some Malaysian forest trees. Philos Trans R Soc Lond B, 335:379-388

Allsopp D, Colwell RR, Hawksworth DL (eds) (1995) Microbial diversity and ecosystem function. CAB International, Wallingford

Anderson JM, Swift MJ (1983) Decomposition in tropical forests. In: Sutton SL, Whitmores TC, Chadwick PC (eds) Tropical rain forest: ecology and management. Blackwell, Oxford, pp 287-309

Anon (1946) *Armillaria mellea* in Nyasaland. Rep Forest Dep Nyasaland 1946: 6

Arnolds E (1991) Mycologist and conservation. In: Hawksworth DL (ed) Frontiers in mycology. CAB International, Wallingford, pp 243-264

Arsuffi TL, Suberkropp K (1984) Leaf processing capabilities of aquatic hyphomycetes: interspecific differences and influence on shredder feeding preferences. Oikos 42:144-154

Ashford WG, Allaway WG (1982) A sheathing mycorrhiza on *Pisonia grandis* R. Br. (Nyctaginaceae) with development of transfer cells rather than Hartig net. New Phytol 90:511-519

Ashford WG, Allaway WG (1985) Transfer cells and Hartig net in the root epidermis of the sheathing mycorrhiza of *Pisonia grandis* R. Br. from Seychelles. New Phytol 100:595-612

Augspurger CK (1983) Seed dispersal of the tropical tree, *Platypodium elegans*, and the escape of its seedlings from fungal pathogens. J Ecol 71:759-771

Baker KB, Cook FJ (1974) Biological control of plant pathogens. Freeman, San Fransisco

Barbosa P, Krischik VA, Jones CG (eds) (1991) Microbial mediation of plant-herbivore interactions. Wiley, New York, 530 pp

Bartlett KB, Harriss RC (1993) Review and assessment of methane emission from wetlands. Chemosphere 26:261-320

Bentley BL (1987) Nitrogen fixation by epiphylls in a tropical rainforest. Ann Mo Bot Gard 74:234-241

Bentley BL, Carpenter EJ (1980) The effects of desiccation and rehydration on nitrogen fixation by epiphylls in a tropical rainforest. Microb Ecol 6:109-113

Bentley BL, Carpenter EJ (1984) Direct transfer of newly-fixed nitrogen from free-living epipyllous microorganisms to their host plant. Oecologia 63:52-56

Berthelin J (1983) Microbial weathering processes. In: Krumbein W. (ed) Microbial geochemistry. Blackwell, Oxford, pp 223-262

Bharat R, Upadhyay RS, Srivastava AK (1988) Utilization of cellulose and gallic acid by litter inhabiting fungi and its possible implication in litter decomposition of a tropical decidous forest. Pedobiologia 32:157-165

Bills G, Polishook J (1994) Abundance and diverstiy of microfungi in leaf litter of a lowland rain forest in Costa Rica. Mycologia 86:187-198

Blair JM, Parmelee RW, Beare MH (1980) Decay rates, nitrogen fluxes, and decomposer communities of single- and mixed-species foliar litter. Ecology 71:1976-1985.

Boddy L (1993) Saprotrophic cord-forming fungi: warfare strategies and other ecological aspects. Mycol Res 97:641-655

Bolland L (1984) *Phellinus noxius*: cause of a significant root-rot in Queensland hoop pine plantations. Aust For 47:2-10

Bonde TA, Christensen BT, Cerri CC (1992) Dynamics of soil organic matter as reflected by natural 13C abundance in particle size fractions of forested and cultivated oxisols. Soil Biol Biochem 24:275-277

Bowen GD (1980) Mycorrhizal roles in tropical plants and ecosystems. In: Mikola P (ed) Tropical mycorrhiza research. Clarendon Press, Oxford pp 165-190

Bravo-Velasquez E, Hedger J (1988) The effect of ecological disturbance on competition between *Crinipellis perniciosa* and other tropical fungi. Proc R Soci Edinb 94B:159-166

Burdon JJ, Chilvers GA (1982) Host density as a factor in plant disease ecology. Annu Rev Phytopathol 20:143-166

Burgess A (1960) Dynamic equilibrium in the soil. In Parkinson D, Ward JS (eds) The ecology of soil fungi : Liverpool Univ Press, Liverpool, pp 185-191

Burghouts TBA, Campbell EJF, Koldermann PJ (1994) Effects of tree species heterogeneity on leaf fall in primary and logged dipterocarp forest in the Ulu Segama Forest Reserve, Sabah, Malaysia. J Trop Ecol 10:1-26

Cabrera, M L (1993) Modelling the flush of nitrogen mineralization caused by drying and rewetting soils. Soil Sci Soc Am J 57:63-66

Castello JD, Leopold DP, Smallidge PJ (1995) Pathogens, patterns, and processes in forest ecosystems. BioScience 45:16-24

Castillo Cabello GP, Georis P, Demoulin V (1994) Salinity and temperature effects on growth of three fungi from Laing Island (Papua New Guinea). Abstr 5th Int Mycol Cong, Vancouver, BC, Canada, August 14-21, 1994. Mycol Soc Am & Int Mycol Soc, 31 pp

Cherrett JM, Powell RJ, Stradling DJ (1989) The mutualism between leaf-cutting ants and their fungus. In: Wilding N, Collins NM, Hammond PM, Webber JF (eds) Insect-fungus interactions. Academic Press, London, pp 93-120

Clark DA, Clark DB (1984) Spacing dynamics of a tropical rain forest tree: evaluation of the Janzen-Connell model. Am Nat 124:769-788

Clay K (1988) Clavicipitaceous fungal endophytes of grasses: coevolution and the change from parasitism to mutualism. In Pirozynski KA, Hawksworth DL (eds) Coevolution of fungi with plants and animals. Academic Press, London, pp 79-105

Collins NM, Sutton SL, Whitmore TC, Chadwick AC (1983) Termite populations and their role in litter removal in Malaysian rain forests. In Sutton SL, Whitmore TC, Chadwick AC (eds) Tropical rain forest: ecology and management. Blackwell, Oxford, pp 311-325

Connell JH (1971) On the role of natural enemies in preventing competitive exclusion in some marine animals and in rain forest trees. In: van der Boer PJ, Gradwell GR (eds) Dynamics of numbers in populations. Proc Adv Study Inst, Ostebeek 1970. Cent Agric Publ Doc, Wageningen, pp 298-312

Cook RJ (1977) Management of the associated microbiota. In: Horsefall JG, Cowling EB (eds) Plant disease - an advanced treatise, vol I. How disease is managed. Academic Press, New York, pp 146-152

Cornejo FJ, Varela A, Wright SJ (1994) Tropical forest litter decomposition under seasonal drought: nutrient release, fungi and bacteria. Oikos 70:183-190

Coxson DS (1991) Nutrient release from epiphytic bryophytes in tropical montane rain forest (Guadeloupe). Can J Bot 69:2122-2129

Curtis H (1975) Biology. 2nd edn. Worth, New York

Davidson EA, Matson PA, Vitousek PM, Riley R (1993) Processes regulating soil emissions of NO and N_2O in a seasonally dry tropical forest. Ecology 74:130-139

Deacon JW, Fleming LV (1992) Interactions of ectomycrrhizal fungi. In: Allen MF (ed) Mycorrhizal functioning, an integrative plant-fungal process. Chapman & Hall, New York, pp 249-300

Dixon RK, Brown S, Houghton RA, Solomon AM, Trexler MC, Wisniewski J (1994) Carbon pools and flux of global forest ecosystems. Science 263:185-190

Dowson CG, Boddy L, Rayner ADM (1989a) Development and extension of mycelial cords in soil at different temperatures and moisture contents. Mycol Res 92:383-391

Dowson CG, Springham P, Rayner ADM, Boddy L (1989b) Resource relationships of foraging mycelial systems of *Phanerochaete velutina* and *Hypholoma fasciculare* in soil. New Phytol 111:501-509

Edgar JG, Kile GA, Almond CA (1976) Tree decline and mortality in selectively logged eucalypt forests in central Victoria. Aust For 39:288-303

Edmisten J (1970) Preliminary studies of the nitrogen budget of a tropical rainforest. In Odum HT, Pigeon RF (eds) A tropical rain forest: a study of irradiation and ecology at El Verde, Puerto Rico. Div Tech Inf, US Atomic Energy Authority, Springfield, Virginia, H211-215

Edmisten JA, Kline JR (1968) Nitrogen fixation by epiphyllae. Puerto Rico Nucl Cent Annu Rep 119:141-143

Egger KN (1995) Molecular analysis of ectomycorrhizal fungal communities. Can J Bot 73 (suppl.): S1415-S1422

Esteban ML, Dorda J, Muller A, Bermudez-de-Castro F (1987) The *Elaeagnus augustifolia* wood at Valdemoro (Madrid). Bol Estac Cent Ecol 16:83-91

Ewel JJ, Mazzarino MJ, Berish CW (1991) Tropical soil fertility changes under monocultures and successional communities of different structure. Ecol Appl 1:289-302

Fagg PC, Ward BK, Featherton GR (1986) Eucalypt dieback associated with *Phytophthora cinnamomi* following logging, wildfire, and favorable rainfall. Aust For 49:36-43

Fellner R (1993) Air pollution and mycorrhizal fungi in central Europe. In: Pegler D, Boddy NL, Ing P, Kirk PM (eds) Fungi of Europe. Royal Botanic Gardens, Kew, pp 239-250

Forman RTT (1975) Canopy lichens with blue-green algae: a nitrogen source in a Colombian rain forest. Ecology 56:1176-1184

Fox RA (1970) The role of biological eradication in root-disease control in replanting of *Hevea brasiliensis*. In: Baker KF, Snyder WC (eds) Ecology of soil-borne plant pathogens. Univ California Press, Berkeley, pp 348-362

Fritz-Sheridan RP, Portecop J (1987) Nitrogen fixation on the tropical volcano, La Soufriere (Guadeloupe). I. A survey of nitrogen fixation by blue-green algal microepiphytes and lichen endophytes. Biotropica 19:194-199

Frobisher M, Hinsdill RD, Crabtree KT, Goodheart CR (1974) Fundamentals of microbiology. 9th edn. Saunders, Philadelphia, 850 pp

Galloway DJ (ed) (1991) Tropical lichens: their systematics, conservation, and ecology. Clarendon Press, Oxford, 302 pp

Gentry AH (1988a) Tree species richness of upper Amazonian forests. Proc Nat Acad Sci USA 85:156 159

Gentry AH (1988b) Changes in plant community diversity and floristic composition of environmental and geographical gradients. Ann Mo Bot Gard 75:1-34

Giavelli G, Bodini A (1990) Plant-ant-fungus communities investigated through qualitative modelling. Oikos 57:357-365

Gilbert GS, Hubbell SP, Forster RB (1994) Density and distance-to-adult effects of a canker disease of trees in a moist tropical forest. Oecologia 98:100-108

Griffin PS, Indictor N, Koestler RJ (1991) The biodeterioration of stone: a review of deterioration mechanisms, conservation case histories and treatment. Int Bioderior 28: 187-207

Haldane JBS (1927) Possible worlds and other essays. Chatto & Windus, London, 312 pp

Hammond PM (1990) Insect abundance and diversity in the Dumoga-Bone National Park, N. Sulawesi, with special reference to the beetle fauna of lowland rain forest in the Toraut region. In: Knight WJ, Holloway JD (eds) Insects and the rain forests of South East Asia (Wallacea). Royal Entomological Society, London, pp197-254

Hammond PM (1992) Species inventory. In: Groombridge B (ed) Global diversity. Chapman & Hall, London, pp 17-39

Harley JL, Smith SE (1983) Mycorrhizal symbiosis. Academic Press, London, 483 pp

Hawksworth DL (1988) Effects of algae and lichen-forming fungi on tropical crops. In: Agnihotri VP, Sarbhoy AK, Kumar D (eds) Perspectives in mycopathology. Malhotra Publishing, New Dehli, pp 76-83

Hawksworth DL (1991) The fungal dimension of biodiversity: magnitude, significance, and conservation. Mycol Res 95:641-655

Hawksworth DL (1993) The tropical fungal biota: census, pertinence, prophylaxis, and prognosis. In: Isaac S, Frankland JC, Watling R, Whalley AJS (eds) Aspects of tropical mycology. Cambridge Univ Press, Cambridge, pp 265-293

Hawksworth DL, Colwell RR (1992) Microbial Diversity 21: biodiversity amongst microorganisms and its relevance. Biodiversity Conserv 1:221-226

Hanksworth DL, Kirk PM,Sutton BC, Pegler DN (1995) Ainsworth & Bisby's Dictionary of Fungi, 8th ed. CAB International, Wallingford

Hawksworth DL, Ritchie JM (1993) Biodiversity and biosystematic priorities: microorganisms and invertebrates. CAB International, Wallingford

Hawksworth DL, Weng Y (1990) Lichens on camphor trees along an air pollution gradient in Hangzhou (Zhejiang Province). For Res 3: 514-517

Hedger, J. (1985) Tropical agarics, resource relations and fruiting periodicity. In: Moore D, Casselton LA, Wood DA, Frankland JC (eds) Developmental biology of higher fungi. Cambridge Univ Press, Cambridge, pp 41-86

Holler JR, Cowley GT (1970) Response of soil, root, and litter microfungal populations to radiation. In: Odum HT, Pigeon RF (eds) A tropical rain forest: a study of irradiation and ecology at El Verde, Puerto Rico. US Atomic Energy Commission. US Dep Commerce, Springfield, VA 22161, pp F 35-39

Hooper DU, Vitousek PM (1992) Biological diversity and terrestrial ecosystem biogeochemistry. In: Schulze E-D, Mooney HA (eds) Biodiversity and ecosystem function. Springer, Berlin Heidelberg New York, pp 3-12

Hueck van der Plas EH (1968) The microbiological deterioration of porous building materials. Int Biodeterior 4:11-28

Hutchison, L.J. (1990) Studies on the systematics of ectomycorrhizal fungi in axenic culture. IV. The effects of some selected fungitoxic compounds upon linear growth. Can J Bot 68:2172-2178

Hutton RS, Rasmussen RA (1970) Microbiological and chemical observations in a tropical forest. In: Odum HT, Pigeon RF (eds) A tropical rain forest: a study of irradiation and ecology at El Verde, Puerto Rico. Oak Ridge, Tenn US Atomic Energy Commision, F43-F56

Hywel-Jones, N. (1993) A systematic survey of insect fungi from natural, tropical forest in Thailand. In: Issac S, Frankland JC, Watling R, Whalley AJS (eds) Aspects of tropical mycology. Cambridge Univ Press, Cambridge, pp 300-301

Ineson PM. Leonard MA, Anderson JM (1982) Effect of collembolan grazing upon nitrogen and cation leaching from decomposing leaf litter. Soil Biol Biochem 14:601-605

Jaenike J (1991) Mass extinction of European fungi. Trends Ecol Evol 6:174-175

Janos DP (1980) Vesicular arbuscular mycorrhiza affect lowland tropical rainforest plant growth. Ecology 62:151-162

Janos, D. P. 1983. Tropical mycorrhizas, nutrient cycles and plant growth. In: Sutton SL, Whitmore TC, Chadwick AC (eds) Tropical rain forest: ecology and management, Blackwell, Oxford, pp 327-345

Janzen DH (1970) Herbivores and the number of trees in tropical forests. Am Nat 104:501-28

Jorgensen RG, Brookes PC, Jenkinson DS (1990). Survival of the soil microbial biomass at elevated temperatures. Soil Biol Biochem 22: 1129-1136

Jun Y, Bridge PD, Evans HC (1991) An integrated approach to the taxonomy of the genus *Verticillium*. J Gen Microbiol 137:1437-1444

Kellas JDG. Kile GA, Jarrett RG, Morgan BJT (1987) The occurrence and effects of *Armillaria luteobubalina* following partial cutting in mixed eucalypt stands in the Wombat Forest, Victoria. Aust For Res 17:263-276

Kelley WD (1987) Effect of tridimefon on development of mycorrhizae from natural inoculum in loblolly pine nursery beds. South J Appl For 11:49-52

Kile GA (1981) *Armillaria luteobubalina*: a primary cause of decline and death of trees in mixed species eucalypt forests in central Victoria, Australia. Aust For Res 11:63-77

Kile GA (1983) Identification of genotypes and the clonal development of *Armillaria luteobubalina* Watling and Kile in eucalypt forests. Aust J Bot 31:657-671

Knops JMH, Nash TH III, Schlesinger WH (1992) The influence of epiphytic lichens on the annual nutrient cycling and on atmospheric deposition of nutrients in an ecosystem. Progr Abstr 77th Ann ESA Meet, August 9-13, 1992, Honolulu, Hawaii. Bull Ecol Soc Am 73: 234

Laessøe T, Lodge, DJ (1994) Three host-specific *Xylaria* species. Mycologia 86:436-446

Last, FT, Natarajan K, Hohan V, Mason PA (1992) Sequences of sheathing (ecto-) mycorrhizal fungi associated with man-made forests, temperate and tropical. In: Read DJ, Lewis DH, Fitter AH, Alexander IJ (eds) Mycorrhizas in ecosystems. CAB International, Oxon, UK, pp 214-219

Leach R (1939) Biological control and ecology of *Armillaria mellea* (Vahl.) Fr Trans Brit Mycol Soc 23:320-329

Lepidi AA, Schippa G (1973) Some aspects of the growth of chemotrophic and heterotrophic microorganisms on calcareous surfaces. In: Romanowski V (ed) 1st Int Symp Deterioration of Building Stone. Imprimeries Reunies, Chamberg, 143 pp

Lodge DJ (1987a) Resurvey of mycorrhizal associations in the El Verde rainforest, Puerto Rico. In: Sylvia DM, Hung LL, Graham JH (eds) Mycorrhiza in the next decade, practical applications and research priorities. Inst Food Agric Sci, Univ Florida, Gainsville, on p. 127

Lodge DJ (1987b) Nutrient concentrations, percentage moisture, and density of field-collected fungal mycelia. Soil Biol Biochem 19:727-733

Lodge DJ (1993) Nutrient cycling by fungi in moist tropical forest. In: Isaac S, Frankland JC, Watling R, Whalley AJS (eds) Aspects of tropical mycology. Cambridge Univ Press, Cambridge, pp 37-57

Lodge DJ, Asbury CE (1988) Basidiomycetes reduce export of organic matter from forest slopes. Mycologia 80:888-890

Lodge DJ, Cantrell S (1995) Fungal communities in wet tropical forests: variation in time and space. Can J Bot 73 (suppl.): S1391-1398

Lodge DJ, McDowell WH, McSwiney CP (1994) The importance of nutrient pulses in tropical forests. Trends Ecol Evol 9:384-387

Lücking R (1992) Zur Verbreitungsökologie foliikoler Flechten in Costa Rica, Zentralamerika. Teil 1-2. Nova Hedwigia 54:309-353; Herzogia 9:181-212

Maggi O, Persiani AM, Casado MA, Pineda FD (1990) Edaphic mycoflora recovery in tropical forests after shifting cultivation. Acta Oecol 11:337-350

Martin MM (1991) The evolution of cellulose digestion in insects. Philos Trans R Soc Lond B 333:281-288

Martin PAW, Travers RS (1989) Worldwide adundance and distribution of *Bacillus thuringiensis* isolates. Appl Environ Microbiol 55:2437-2442

Mathews E, Fung I (1987) Methane emission from natural wetlands: global distributions, area and environmental characteristics of sources. Global Biogeochem Cycles 1:61-86

Matson PA, Vitousek PM (1981) Nitrification potentials following clearcutting in the Hoosier National Forest, Indiana. For Sci 27:781-791

Matson PA, Vitousek PM (1990) Ecosystem approach for development of a global nitrous oxide budget. BioScience 40:667-672

Miller RM (1989) The ecology of vesicular-arbuscular mycorrhizae in grass- and shrublands. In: Safir GR (ed) Ecophysiology of VA mycorrhizal plants. CRC Press, Boca Raton, pp 135-158

Nash TH III, Wirth V (eds) (1988) Lichens, bryophytes and air quality. [Bibliotheca Lichenologica No 30.] J. Cramer, Berlin, 297 pp

Newell K (1984) Interaction between two decomposer basidiomycetes and a collembolan under Sitka spruce: grazing and its potential effect on fungal distribution and litter decomposition. Soil Biol Biochem. 16:235-239

Olsen JS (1963) Energy storage and the balance of producers and decomposers in ecological sytems. Ecology 44:322-331

Ortega-Calvo JJ, Hernandez-Marine M, Saiz-Jimenez C (1991) Biodeterioration of building materials by cyanobacteria and algae. Int Biodeterior 28:165-185

Parker GS (1994) Soil fertility, nutrient acquisition, and nutrient cycling. In: McDade LA, Bawa KS, Hespenheide HA, Hartshorn GS (eds) La Selva. Ecology and natural history of a neotropical rain forest. Univ Chicago Press, Chicago, pp 54-63

Payne W (1973) Reduction of nitrogenous oxides by micro-organisms. Microbiol Rev 37:409-452

Perry N (1990) Symbiosis. Nature in partnership. Blandford Press, London, 128 pp

Pichel RJ (1956) Root rots of *Hevea* in the Congo basin. Inst Nat Etude Agron Congo belge Ser Tech 49:480

Powell C (1980) Mycorrhizal infectivity of eroded soils. Soil Biol Biochem 12:247-250

Price PW (1988) An overview of organismal interactions in ecosystems in evolutionary and ecological time. Agric Ecosyst Environ 24:369-377

Prinn R, Cunnold D, Rasmussen R, Simmonds P, Aleya F, Crawford P, Fraser R, Rosen R (1990) Atmospheric emissions and trends of nitrous oxide deduced from 10 years of ALE-GAGE data. J Geophys Res 95(D 11):18:369-385

Raghubanshi AS, Srivastava SC, Singh RS, Singh JS (1990) Nutrient release in leaf litter. Nature 346:227

Redfern DB (1975) The influence of food base on rhizomorph growth and pathogenicity of *Armillaria mellea* isolates. In: Bruehl GW (ed) Biology and control of soil-borne plant pathogens. The American Phytopathological Society, St Paul, MN, pp 69-73

Redhead JF (1980) Mycorrhiza in natural tropical forests. In: Mikola P (ed) Tropical mycorrhiza research. Clarendon Press, Oxford, pp 127-142

Richardson DH (ed) (1987) Biological indicators of pollution. Royal Irish Academy, Dublin, 242 pp

Richardson DH (1992) Pollution monitoring with lichens. Richmond, Slough, Naturalists' Handb 19:76

Rishbeth J (1955) Root diseases in plantations, with special reference to tropical crops. Ann Appl Biol 42:220-227

Rishbeth J (1980) *Armillaria* on cacao in Sao Tome. Trop Agric Trinidad 57:155-165

Robertson GP, Tiedje JM (1988) Deforestation alters denitrification in a lowland tropical rain forest. Nature 336:756-759

Rodrigues KA, Leuchtmann A, Petrini O (1993) Endophytic species of *Xylaria*: cultural and isozymic studies. Sydowia 45:116-138

Rose F (1992) Temperate forest management: its effects on bryophyte and lichen floras and habitats. In: Bates JW, Farmer AM (eds) Bryophytes and lichens in a changing environment. Clarendon Press, Oxford, pp 211-233

Rose SL, Trappe JM (1980) Three new endomycorrhizal *Glomus* spp. associated with actinorrhizal shrubs. Mycotaxon 10:413-420

Rose SL, Youngberg CT (1981) Tripartite associations in snowbrush (*Ceanthus velutinus)*: effect of vesicular arbuscular mycorrhizae on growth nodulation and nitrogen fixation. Can J Bot 59:34-39

Rossman AY (1994) The need for identification services in agriculture. In: Hawksworth DL (ed) The identification and characterization of pest organisms CAB International, Wallingford, pp 35-45

Sanchez P, Gichuru MP, Katz LB (1982) Organic matter in major soils of the tropical and temperate regions. Trans 12th Int Congr Soil Sci New Delhi 1:99-114

Schmitz MF, Aranda Y, Esteban ML, Bermudez de Castro F (1990) Nodulation of *Elaeagnus angustifolia* in the wood at Valdemoro (Madrid). Ecol Madrid 4:121-129

Seaward MRD (1988) Contribution of lichens to ecosystems. In: Galun M (ed) CRC handbook of lichenology, vol 2. CRC Press, Boca Raton, pp 107-129

Seiler W, Conrad R, Scharffe D (1984) Field studies of methane emission from termite nests into the atmosphere and measurements of methane uptake by tropical soils. J Atmos Chem 1:171-186

Sheriden RP (1991) Epicaulous, nitrogen-fixing microepiphytes in a tropical mangal community, Guadaloupe, French West Indies. Biotropica 23:314-322

Smith ML, Bruhn JN, Anderson JB (1992) The fungus *Armillaria bulbosa* is amongst the largest and oldest living organisms. Nature 356:428-431

Srivastava SC (1992) Microbial C, N and P in dry tropical soils: seasonal changes and influence of soil moisture. Soil Biol Biochem 24:711-714

Swift MJ, Anderson JM (1989) Decomposition. In: Leith H, Werger MJA (eds) Tropical rain forest ecosystems. Elsevier, Amsterdam, pp 547-569

Swift MJ, Heal OW, Anderson MJ (1979) Decomposition in terrestrial ecosystems. Blackwell, Oxford, 372 pp

Tacon F le, Bouchard D, Perrin R (1986) Effects of soil fumigation and inoculation with pure culture of *Hebeloma cylindrosporum* on survival, growth and ectomycorrhizal development of Norway spruce and Douglas fir seedlings. Eur J For 16:257-265

Tacon F le, Garbaye J, Beddiar AF, Diagne O, Diem HG (1989) The importance of root symbiosis for forest trees in dry and wet tropical forests. Trees for development in Sub-Saharan Africa. Proc Regional Sem held by the International Foundation for Science (IFS), Nairobi, pp 302-318

Thoen D (1993) Looking for ectomycorrhizal trees and ectomycorrhizal fungi in tropical Africa. In: Isaac S, Frankland JC, Watling R, Whalley AJS (eds) Aspects of tropical mycology. Cambridge Univ Press, Cambridge, pp 193-205

Thompson W, Boddy L (1983) Decomposition of suppressed oak trees in even-aged plantations. II. Colonization of tree roots by cord and rhizomorph producing basidiomycetes. New Phytol 93:227-291

Thrower SL (1980) Air pollution and lichens in Hong Kong. Lichenologist 12:305-311

Torrey JG (1982) *Casuarina*: actinorrhizal nitrogen-fixing tree of the tropics. In: Graham PH, Harris SC (eds) Biological nitrogen fixation technology for tropical agriculture. CIAT, Colombia, pp 427-439

Trüper HG (1992) Prokaryotes: an overview with respect to biodiversity and environmental importance. Biodiv Conserv 1:227-236

Turner PD (1991) *Ganoderma* in Oil Palm: a review of the current situation (1991) in Malaysia and Indonesia. Harrison Fleming Advisory Services, Duns, 20 pp

Unestam T, Chakravarty P, Damm E (1990) Fungicides: in vitro tests not useful for evaluating effects on mycorrhizae. Agric Ecosyst Environ 28:535-538

Villarreal L, Perez-Moreno J (1989) Wild edible fungi from Mexico, an integral approach. Micol Neotrop Apl 2:77-114

Vitousek PM (1990) Biological invasions and ecosystem processes: towards an integration of population biology and ecosystem studies. Okios 57:7-13.

Vitousek PM, Hooper DU (1993) Biological diversity and terrestrial ecosystem biogeochemistry. In: Schulze E-D, Mooney HA (eds) Biodiversity and ecosystem function. Springer, Berlin Heidelberg New York, pp 3-14

Vitousek PM, Walker LR (1989) Biological invasion by *Myrica faya* in Hawaii: plant demography, nitrogen fixation, and ecosystem effects. Ecol Monogr 59:247-265

Weaver PL (1972) Cloud moisture interception in the Luquillo Mountains of Puerto Rico. Carib J Sci 12:129-144

Wells JM, Boddy L (1990) Wood decay, and phosphorus and fungal biomass allocation, in mycelial cord systems. New Phytol 116:285-295

Weste G (1986) Vegetation changes associated with invasion by *Phytophthora cinnamomi* of defined plots in the Brisbane Ranges, Victoria, 1975-1985. Aust J Bot 34:633-648

Weste G, Marks GC (1987) The biology of *Phytophthora cinnamomi* in Australasian forests. Annu Rev Phytopathol 25:207-229

Wheeler Q, Blackwell M (eds) (1984) Fungus-insect relationships. Columbia Univ Press, New York, 514 pp

Whitton BA (1992) Diversity, ecology, and taxonomy of the cyanobacteria. In: Mann NH, Carr NG (eds) Photosynthetic prokaryotes. Plenum Press, New York, pp 1-51

Wilding N, Collins NM, Hammond PM, Webber JF (eds) (1989) Insect-fungus interactions. Academic Press, London, 344 pp

Willig MR, Camilo GR (1991) The effect of Hurricane Hugo on six invertebrate species in the Luquillo Experimental Forest of Puerto Rico. Biotropica 23:455-461

Wilson J, Mason PA (1993). Population dynamics of vesicular-arbuscular mycorrhizal (VAM) fungi in secondary moist forest in West Africa. In: Isaac S, Frankland JC, Watling R, Whalley AJS (eds) Aspects of tropical mycology. Cambridge Univ Press, Cambridge, 299 pp

Wolseley PA, Aguirre-Hudson B (1991) Lichens as indicators of environmental change in the tropical forests of Thailand. Global Ecol Biogeogr Lett 1:170-175

Wood TG, Thomas RJ (1989) The mutualistic association between Macrotermitinae and *Termitomyces*. In: Wilding N, Collins M, Hammond PM, Webber JF (eds) Insect-fungus interactions. Academic Press, London, pp 62-92

Yang J, Insam H (1991) Microbial biomass and relative contributions of bacteria and fungi in soil beneath tropical rain forest, Hainan Island, China. J Trop Ecol 7:385-393

Young A (1976) Tropical soils and soil survey. Cambridge Univ Press, Cambridge

Zimmerman JK, Pulliam WM, Lodge DJ, Quinones V, Fetcher N, Guzman-Grajales S, Waide RB, Parrotta JA, Asbury CE, Walker LR (1995) Decomposition of coarse woody debris limits short-term recovery from hurricane damage by subtropical wet forest in Puerto Rico. Oikos 72:314-322

6 Plant Life-Forms and Tropical Ecosystem Functioning

John J. Ewel[1] and Seth W. Bigelow[1]

Structure without function is a corpse; function sans structure is a ghost.
(Vogel 1972)

6.1 Introduction

Life-form was said by Warming (1909) to represent the sum of adaptive characters in a species, and thus is an expression of the harmony between a plant and its environment. This colorful perspective, although too broad to be useful in classification, does highlight the essential point that life-form groupings should be ecologically relevant.

With widely accepted phylogenetic classification schemes available, why even seek broad schemes of categorization in which the members of a group share no evolutionary history? The answer, simply put, is that form dictates function, and we contend that, among adults of higher plants adapted to a particular habitat, the smallest unit that exerts major control on mesoscale ecosystem processes is the life-form. Many higher plants are functionally equivalent: they all consume carbon dioxide and water; they all use solar energy of the same wavelengths; and they all require the same suite of 13 mineral nutrients. To be sure, there are some differences (the pigments of some algae have different absorption spectra, some congeners specialize on different chemical species of nitrogen, some plants are symbiotic with microorganisms capable of fixing diatomic nitrogen), but these describe the exceptions: for the most part, all plants feed out of the same trough.

Now the heresy: preoccupation with the consequences of loss of plant species on ecosystem functioning (energy and material fluxes, for example) is probably unwarranted. If, among higher plant species, there is substantial redundancy in nature, then loss of a minor player or substitution of one species for another is likely to have immeasurably small consequences for mesoscale processes such as carbon and oxygen exchange, soil erosion, and water and nutrient budgets. But this is not likely to be the case with loss of life-forms, for if there are significant functional differences among plants, surely these will manifest themselves most dramatically among species of

[1] Department of Botany, University of Florida, Gainsville Florida 32611, USA

Ecological Studies, Vol. 122
Orians, Dirzo and Cushman (eds) Biodiversity and Ecosystem Processes in Tropical Forests
© Springer-Verlag Berlin Heidelberg 1996

strikingly different architecture. It is the assemblage of life-forms that give forests their characteristic structure, and structure, in turn, dictates whole-system functioning.

6.1.1 Functional Significance of Life-Forms

Solbrig (1993) accurately described life-forms as single-character-based functional groups, yet even at this level there are some similarities in life history and resource use that lend coherence to the categories. This is per-haps to be expected if life-forms are the result of evolutionary forces that lead to ecological convergence (Böcher 1977). For example, a restricted set of pollen vectors may be available to plants that flower in any given part of the canopy, eventually leading to convergence on a particular set of breed-ing mechanisms.

In fact, several studies have shown good correlation between life-form and various life-history traits in tropical forests. Examples: (1) In a study of a Mexican dry forest, Bullock (1985) found that monostylous hermaphro-ditism was strongly associated with epiphytic and herbaceous life-forms, but less so with trees; (2) within the large, mostly tropical family Rubiaceae, genera that combine particular dispersal systems and life-forms (gravity dispersal in herbs, animal-mediated dispersal in shrubs, and winged seeds in shrubs and trees) tend to have large numbers of species (Eriksson and Bremer 1991); and (3) across a range of forest sites in Venezuela, trees have significantly higher seed, fruit, and flower weight; rates of abortion; and ratios of fruit to flower mass than other life-forms (Ramírez 1993). Thus, even life-form groups as inclusive as "tree" or "herb" may share more char-acters than is generally appreciated.

6.1.2 Assessing the Consequences of Life-Form Diversity

There are only three ways to predict accurately the consequences of loss of life-form diversity; for convenience we dub them the experimental, natural history, and examination of pattern approaches. The most direct method would consist of performing removal (or perhaps addition) experiments, accompanied by measurement of responses on replicated treatment and control plots. Unfortunately, few such experiments have been performed, and few are underway (e.g., Haggar and Ewel 1994). There are, nevertheless, a number of places where an "experiment" has been performed without benefit of replication, controls, or monitoring of response variables. For example, climber cutting is a common silvicultural prescription in Malaysia (e.g., Fox 1968; Putz 1985), and the Kekchí Maya people have killed (in harvesting its fruits) the dominant understory palm (*Astrocaryum*) throughout much of the eastern Guatemalan lowlands. Some of these situations, though not ideal, might lend themselves to a posteriori analysis.

The second approach consists of using a sound knowledge of natural history to deduce relationships and dependencies among life-forms, and to predict the consequences of life-form loss. The primary effects will undoubtedly be easier to deduce than the secondary effects, for example, loss of red-flowered herbs, probably loss of hummingbirds, and maybe loss of snakes that feed on hummingbirds. But what about the impacts on the peccary that feeds on that snake and 43 other things as well? And what effects might this ultimately have on soil churning and erosion? Could any scholar ever acquire the experience and breadth of knowledge required to make such predictions accurately? Not likely, but the potency of the natural history approach should not be underestimated, for it is our only link to the complexity of the real world and our only check on the models and generalizations of theoreticians.

The third approach consists of examining patterns of life-form distribution across the landscape and seeking differences in ecosystem functioning that accompany those patterns. Such an approach will never reveal cause and effect, except circumstantially, but it is the path most readily amenable to analysis using the plethora of vegetation descriptions available from throughout the tropics. We do not attempt the analysis herein, but we do describe some of the players and patterns.

6.2 Classification

The scheme of Christian Raunkiaer, the Danish phytogeographer whose name is synonymous with life-form, is of little help in the tropics, for his system of classifying plants according to the height of over-wintering buds compresses tropical plants into relatively few categories (Raunkiaer 1934). For example, Raunkiaer's system fails to discriminate between two of the most distinct growth habits, self-supporting and climbing. Life-form classification schemes are beset with two classes of problems, both of which impede their use when seeking functional correlates. The first are those that overemphasize a particular trait, often one that is extremely important in the region where the system was developed. Raunkiaer's emphasis on position of the perennating organ is an example: it works well in the Alps and poorly in the Amazon. The other is the tendency to subdivide major groups so finely that the system contains vast numbers of categories, and a manual is required to distinguish one from another; such systems simply do not get used.

Several life-form classifications preserve essential features of Raunkiaer's system while incorporating additional detail likely to be of ecological relevance. One such system, which has the added advantage of having been devised by ecologists exceptionally knowledgeable about the tropics, is the scheme of Mueller-Dombois and Ellenberg (1974). The first tier in their three-layered hierarchy is based on trophic status, the second on vascularization, and the third on support structures; the result is a manageable 23-category system.

Nevertheless, there is no universal agreement on one system of life-form classification. A minimal description of tropical forest plant life-forms would have to include trees, herbs, epiphytes, and climbers, but this would hardly do justice to the astonishing proliferation of morphologies in most equatorial forests. To be useful, a system would have to include at least the seven most-conspicuous categories: dicotyledonous trees; trees with one or few meristems (e.g., arborescent palms and pandans); treelets; shrubs; giant herbs; vines; and epiphytes. For some purposes one might want more classes (forbs, grasses, and succulents would be essential additions in many tropical environments), but in no case would it be useful to compress these into fewer categories.

Tropical trees are, of course, the background fabric of forest structure, and their great importance and diversity have led to several attempts to make ecologically meaningful classifications based on form. These range from simple stature-based systems such as the treelet-understory-sub-canopy-canopy trees of Hartshorn and Poveda (1983) to the elaborate system of models proposed by Hallé et al. (1978). Although stature-based classification systems may have some ecological significance, the same is not necessarily true of the models of Hallé et al. Studies of crown formation have shown that trees are so plastic that different tree architectures can converge on the same crown morphology; conversely, a given model can give rise to quite different crown forms (Fisher 1986). It is during the regeneration phase, when architecture is most clearly expressed, that the models prove their utility. Kohyama (1987), for example, found that architectural models were correlated with crown allometry in saplings, and that these characteristics were sound predictors of performance in gaps.

Environment is, without doubt, the primary driver of life-form evolution. Nevertheless, the very core of every definition of ecology implies two-way interactions between organisms and environment, and just as life-form is a response to environment, the entire *oikos* is influenced by structure. Thus, the product of an evolutionary response to environment becomes a major actor in the interplay among biota, soil, and atmosphere. Two form-linked traits, size (which influences many aspects of resource acquistion) and life span (which affects community turnover), may have special significance.

6.2.1 Stature

Plant size is not an inevitable indicator of resource dominance, for it is the leaves that capture radiation and the mycotrophic roots that retrieve nutrients from the soil, regardless of the massiveness of organs such as stems, buttresses, and support roots. Some life-forms invest heavily in resource-capturing organs (e.g., foliage of some grasses, fine roots of some vines) despite their relatively modest stature overall. Nevertheless, size-class distinctions are important: within a given life form, a plant's ability to gather resources is, in large part, determined by its size. A canopy tree, for example, typically intercepts a far greater proportion of incident light than a sapling; furthermore, this advantage should allow the tree to allocate substantial photosynthate to roots, thus enabling it to explore a greater volume of soil than its shorter neighbors. Domination of the resource base may then influence other life-history traits; Rockwood (1985), for example, examined herbarium specimens from Costa Rica and Panama and found that mean seed weight increased from herbs to shrubs to trees.

The importance of stature, and the functional attributes it imparts, is perhaps best epitomized in the dichotomy between trees and shrubs. Evidence suggests that shrubs and trees are best treated as distinct life-forms and not just part of a continuum that is arbitrarily divided on the basis of size. Compared with trees, shrubs may have proportionally fewer thick, structural roots; a greater fraction of their roots concentrated near the surface; persistent, reproductively active shoots close to the ground; greater ability to reproduce vegetatively (by layering or rooting of crushed or broken branches); and the capacity to return to reproductive mode more quickly following crown damage. Excavation of shrubs and treelets on Barro Colorado Island, Panama, for example, revealed that shrubs have a larger ratio of root-surface area to leaf-surface area than saplings of co-occurring trees (Becker and Castillo 1990). Because shrubs invest in flowers and fruits (which can be major nutrient sinks), whereas saplings of equal height do not, Becker and Castillo hypothesized that shrubs allocate proportionately more carbon to shallow roots. The fact that shrubs have limited need for the thick, belowground anchors required to keep massive trees upright might also contribute to their concentration of roots near the surface.

Species with relatively small lateral branches arranged around a stout trunk grow larger than species with heftier lateral branches; in fact, shrub-like architectures may not even be possible for saplings that will later grow tall (Stevens and Perkins 1992). Most trees shed their lower branches once they no longer make net contributions to carbon balance, but these temporary branches are often relatively short; presumably the structural investment required to produce long branches (thus, a broad crown) is unwarranted unless limbs are to be retained for a long time. Shrubs, in contrast, suffer no future penalty for investing in persistent branches, and they may be more adapted to dominating substantial lateral space relative to their height.

These arguments hold only for dicotyledonous plants, but monocots of all sizes are also prevalent throughout the tropics. Indeed, one reason for the success of arborescent palms may be that they avoid the tradeoff: their unique leaf morphology allows a wide crown on short individuals, without an investment in branches that will later become a permanent metabolic drain when they become shaded. The fiber-rich tissues of broad, arching palm leaves also impart strength and elasticity which enable them to survive branch falls that would fracture most understory dicots.

Studies of the neotropical understory palm genus *Geonoma* have explored the tradeoffs involved in increasing stature to improve light interception (Chazdon 1991). Low-stature plants of the tropical forest understory have to cope with extremely low light levels, but small increases in stature can lead to greatly increased light interception (Chazdon and Fetcher 1984). It might be expected, therefore, that natural selection would inevitably favor height. Nevertheless, Chazdon (1985) demonstrated that increases in stature are linked to increases in leaf size, number, and morphological complexity; but as these factors increase, efficiency of light interception per unit of biomass decreases. Small plants, therefore, have reduced metabolic costs and can grow and reproduce in more shaded conditions. Thus form, stature, and reproductive status are tightly integrated to determine shade tolerance. By foregoing the ability to grow large, the understory palms effectively exploit the scant light resources available in the understory. The opportunity presented by this habitat is reflected in the group's great diversity and, in some tropical forests, its extraordinarily high density (Kahn 1986).

6.2.2 Longevity

Just as it is useful to break life-forms down into stature classes, it is equally useful to distinguish among members of a life-form that have different life spans. Stand turnover has, in the last 20 years, become recognized as a major controlling agent of species richness (Connell 1978, 1979), resource patchiness (Vitousek 1985), and vegetation pattern (Hubbell 1979; Hubbell and Foster 1983; Weinstein and Shugart 1983). Thanks to the gap-dynamic paradigm, ecologists have finally abandoned their visions of tropical forests as static entities comprised of uniformly ancient trees; instead, most forests are recognized as mosaics comprised of unequal-aged patches, as was suggested decades ago by Aubréville (1938); and if different locales are subjected to disturbance with different frequencies, it is not surprising that the species within each life-form are characterized by many different life spans, making longevity a useful second-tier categorization tool.

Plants, which (unlike some animals) do not transfer information to the next generation except through their DNA, pay a fitness penalty for surviving longer than the reproductive phase (because of occupancy of resources by the parent), just as they pay a penalty for dying early, before their reproductive potential is realized. One would expect, therefore, the frequency of longevity classes within a local flora to reflect the dominant frequencies of disturbance. If stand turnover were caused by a single factor of predictable long-term frequency, for example, one might expect all plants in an ecosystem to converge on a single life span; likewise, if two forces acted at different frequencies (fire every 20 years and typhoons every 160, for example), one might expect a bimodal distribution of longevity classes, with one class vulnerable to fire and another class capable of with-standing fire but vulnerable to wind. Tolerance to both fire and wind would presumably lead eventually to dominance by one or more still-longer-lived species, according to the inhibition model of succession (Connell and Slatyer 1977).

Such an accommodating relationship between disturbance regimen and life span does not, of course, obtain in nature, where death is probabilistic and comes from many agents. Furthermore, not all disturbances are exogenous, and those that are endogenous expose different life forms to different frequencies of mortality within the same forest. For example, fallen branches kill understory trees (Clark and Clark 1991) and eventually the epiphytes they carry with them (Matelson et al. 1993), whereas some palms (de Castro y Santos 1980) and shrubs survive.

The situation is further complicated because the information required to enable ecologists to analyze plant longevity in tropical forests is hard to obtain due to the difficulty of aging most tropical perennials (Bormann and Berlyn 1981). The only sure option currently available is long-term obser-vation, a grim prospect when one recognizes that mean life spans of some suites of species are likely to be measured in centuries. Nevertheless, the extremes are apparent: some trees live little more than a decade, whereas others are known (e.g., trees long used as African village gathering points; prominent landmarks) to live for centuries. Curiously, some of the fast-growing, low-wood-density pioneer trees (e.g., several Bombacaceae and Dipterocarpaceae) are also long-lived species.

Because the direct aging of long-lived organisms that yield no anatomical clues is extremely difficult, it might be useful to attempt to estimate longevity classes indirectly, using another variable. A good candidate might be growth rate, for there is some evidence, both empirical (see discussion in Ewel 1986) and theoretical (Pearl 1928; Rose 1991), that within a species, individuals with the fastest growth rates come to the end of their allotted life span soonest.

Nevertheless, some life-forms do not lend themselves to such an approach because they couple short-lived (often fast-growing) shoots with an almost irrepressible ability to resprout from a long-lived base. Examples include large woody vines (shoots of which usually die when they tumble out of the canopy) and understory herbs (whose shoots frequently succumb to branch fall). Ecosystem functioning might be affected in one way by the longevity of the ramet and in another by the longevity of the genet, and in such cases it would be useful to make a distinction between the two longevity classes of the same "individual."

One otherwise conspicuous longevity class does seem to be missing, or nearly so, from humid tropical forests: annuals. These short-lived plants are creatures of harsh season and high frequency of disturbance; where they abound in the humid tropics they seem to have tracked human activity. Coincidentally, they also tend to have higher relative growth rates than most longer-lived plants (e.g., Grime and Hunt 1975).

6.3 Biogeographical Patterns

There is undeniably a huge increase in species diversity along a gradient from the poles towards the equator, but whether there is a concomitant increase in life-form diversity is debatable. Clearly, life-form diversity depends on the classification system used, but what should be expected in theory? Plants in the tropics operate under relaxed environmental constraints, and the lack of a cold winter may have permitted the proliferation of plant morphologies that would have been selected against in harsher climes. Tree architectural diversity (sensu Hallé et al. 1978) reaches its zenith in the tropics, as does the life-form diversity of epiphytes (Gentry 1988). Although this augurs for higher life-form diversity in the tropics, according to Box (1981) tropical regions contain 18 life-forms, whereas the subtropics have 20, and temperate regions no fewer than 25. Raunkiaer predicted that life-form diversity should be highest in the mid-latitudes where the over-wintering season is intermediate in harshness, and his prediction seems to be borne out by Box' analysis.

Life-form spectra are by no means identical in tropical forests on different continents, even in areas of comparable climate. Having sampled some 80 sites, for example, Gentry (1991) reported that climbers (and hemiepiphytes) were more abundant in Africa (> 1000 stems per hectare) than in the Neotropics or Australasia (about 700/ha in each). These data, while intriguing, are based on modest sample sizes (11 plots of 1000 m^2 in continental Africa and Madagascar, 56 in the Neotropics, and 13 in Australasia), and more extensive documentation would be desirable. The distribution and abundance of vines is particularly complex because so many families are involved and because vine abundance is inextricably confounded with site history and climate.

The relative paucity of arborescent palms in Africa and Asia compared to the Neotropics is better documented (Moore 1973), as one might expect easy-to-count members of a single, conspicuous family. Across a broad range of lowland rain forest sites in the Neotropics, arborescent palms accounted for roughly 15% of the trees, whereas they comprised less than 2% of the adult trees in four African sites and were completely absent from two Asian sites (Grubb et al. 1964; Emmons and Gentry 1983).

Island floras are notoriously depauperate. Are they also disharmonic (sensu MacArthur and Wilson 1967), i.e., do they lack many of the adaptive types found on their donor continents? It might be reasonable to predict that life-forms with predominantly wind-dispersed seeds (e.g., vines, as demonstrated by Gentry 1983, 1991) or extremely large, mammal-dispersed seeds (e.g., many trees and arborescent palms) would be under-represented in island floras, whereas those with bird-dispersed seeds (e.g., shrubs) would be disproportionately abundant. On the other hand, the life-form spectrum of the Hawaiian Islands is not skewed relative to that of continents, even though the Hawaiian flora is species-poor overall (Mueller-Dombois et al. 1981). Comparisons of islands having disharmonic floras with continents would be potent tools for elucidating relationships between life-form diversity and ecosystem functioning.

6.4 Environmental Correlates of Life-Form Diversity

Life-form diversity is correlated with a number of abiotic variables. The most elaborate attempt in recent years to predict life-form distribution as a function of climate was that of Box (1981), who used six climatic variables to define the niche space of each of 77 life-forms. His tropical life-forms include such categories as Tropical Linear-Leaved Trees, Tropical Broad-Evergreen Small Trees, and Tropical Broad-Evergreen Lianas. The impressive agreement between actual and predicted occurrence (based on Box' model) indicates that, at a global scale, life-form distribution is largely controlled by climate.

6.4.1 Rainfall

Within the tropics, precipitation is a major determinant of life-form distribution (Table 6.1). If life-form definitions emphasize the types most important in the tropics, the general trend is one of greater life-form richness with increasing rainfall. If, on the other hand, Raunkiaer's system, with its emphasis on adaptations to harsh environments, is used, life-form richness increases with aridity (e.g., Shreve 1936; Whittaker and Niering 1965).

Different life-forms respond in different ways to increasing availability of water, even in tropical forests. Epiphytes are more affected than any other life-form group, and there is a strong correlation between annual precipitation and contribution of epiphytes to species richness (Table 6.1). Along a gradient from dry forest to wet forest in Ecuador, for example, the abundance of epiphytes increased by a factor of 450 and the number of species increased from 2 to 35 (Gentry and Dodson 1987b). Palms also become increasingly dominant (at least in the Neotropics) as precipitation increases (Table 6.1).

John J. Ewel and Seth W. Bigelow

Table 6.1. Life-form composition of selected tropical forests, arranged in order of increasing rainfall. All sites are at elevations lower than 500 m except Mt. Kerigomna, Papua New Guinea, at 2700 m. Values are percentages of species. Blanks indicate that information could not be extracted from data as published

Site Reference (annual rainfall, mm)	Reference	Trees >10 cm DBH	Arborescent palms	Small trees, large shrubs	Terrstrial herbs, small shrubs	Climbers	Epiphytes	Other
Chamela, Mexico (784)	1	26	0	23	24	22	4	
Capeira, Ecuador (804)	2	15		6	52	24	2	1
Ibadan Nigeria (1220)	3	51		16		32	1	
Santa Rosa, Costa Rica (1550)	4	21	0.3	10	48	17	4	1
Makokou, Gabon (1755)	5	34	0.4	37		23	6	1
Jauneche, Ecuador (1855)	6	20		11	36	22	11	1
Rio Manu, Peru (2000)	7	27	0.5	28	14	18	12	
Queensland, Australia (2330)	8	68		10		11	4	7
Barro Colorado, Panama (2750)	9	21	1	10	33	20	13	1
Rio Palenque, Ecuador (2980)	10	15		10	36	16	22	1
La Selva, Costa Rica (4000)	11	21	1	42		12	25	1
Horquetas, Costa Rica (4000)	12	31	1	6	13	24	25	
Mt. Kerigomna, Papua New Guinea (4000)	13	39	0	14	14	16	16	

1 Bullock (1985); 2,4,5,6,9,10,11; Gentry and Dodson (1987a,b); 3 Hall and Okali (1978); 7 Foster (1990); 8 Stocker (1988); 12 Whitmore et al. (1985); 13 Grubb and Stevens (1985)

For example, Santa Rosa, Costa Rica, (1.6 m mean annual precipitation) has 3 palm species, compared to 18 at La Selva, Costa Rica (4 m). Many palms tolerate anoxic conditions well and become the sole dominants on flooded soils (Myers 1990). The large water storage capacity in the trunks of arborescent palms may play a significant role in enabling survival during times when roots can make little contribution to plant needs (Holbrook and Sinclair 1992). In contrast, the contribution of climbers to species richness may decline slightly as precipitation increases (Table 6.1), and herb richness is often greater in dry forests than in wet forests, perhaps owing to the more open canopy (Gentry and Dodson 1987a).

6.4.2 Altitude

The changes in life-form (as well as forest stature and species composition) that accompany change in elevation are readily observed but hard to assess, in part, at least, because three environmental factors – temperature, rainfall, and mist – tend to change simultaneously with altitude. In the humid tropics, temperature drops more or less predictably, at a rate of 5.5 to 6.0 °C per 1000 m. Atmospheric moisture, on the other hand, is more complex. Orographic uplift of incoming air masses often results in an increase in rainfall with elevation. Furthermore, at the lifting condensation level (often in the range of 600 to 2000 m), forests are bathed in mist. Above the lifting condensation level, once the atmosphere has been sapped of its water, both rainfall and humidity decline, so low temperatures often co-occur with aridity at high elevations.

Epiphytes, both vascular and nonvascular, commonly reach their greatest abundance and diversity in montane cloud forest (Gradstein and Pócs 1989). Along an altitudinal transect in Ecuador, for example, Grubb et al. (1964) found 2555 epiphytes in a 465 m^2 montane plot, compared to only 675 individuals in a similarly sized plot in lowland rain forest. Palm diversity declines with elevation, but many mid- to high-elevation forests contain one (or a few) palms in great abundance. In fact, the tallest palm in the world, *Ceroxylon andicola*, which grows to 60 m, is found as high as 3000 m in the Colombian Andes (Cuatrecasas 1958). Climbers appear to maintain an important place in the biological spectrum, at least up to the transition from upper montane forest to cloud forest (Leigh 1975; Grubb 1977). Nevertheless, there is a shift in relative abundances of different types of climbers with altitude in the Neotropics. Gentry (1988), for example, found that hemiepiphytes reached peak abundance at mid-elevations in the Andes, to be supplanted at lower and higher elevations by free-climbing lianas.

6.4.3 Soil Fertility

Climate is not the only environmental factor that influences life-form
distribution; the ability of a soil to meet nutritional needs seems to control
distribution in some cases. In general, fertile soils support a greater profu-
sion of forms than do impoverished soils in the same climate.

Based on his own extensive surveys and data in the literature, Gentry
(1991) concluded that "there is a very slight tendency for greater liana density on richer soils," but within small areas the relationship seems more
pronounced (e.g., Grubb and Tanner 1976; Proctor et al. 1983; Putz 1983;
Putz and Chai 1987). The giant, perennial herbs (e.g., members of the Heliconiaceae, Zingiberaceae, Marantaceae, Musaceae, etc.) so typical of
tropical forests seem to be more abundant and diverse on fertile soils, but
data are scarce, as most authors do not segregate them as a life-form. Epiphytes might seem to be nutritionally autonomous, but tank epiphytes
(and probably others) depend on inputs of detritus, the quality of which is
determined, in part, by soil fertility. Shrubs and understory palm-like
plants (Palmae, Pandanaceae, Cyclanthaceae, and similar forms) are more
abundant on fertile soils, but whether this is a direct effect of nutrition or
an indirect effect of the fast turnover of trees on fertile soils is unknown.

6.5 Episodic Impacts on Life-Form Diversity

In addition to the chronic environmental controls of life-form diversity in
tropical forests, such as climate and soils, a number of agents intervene
locally or aperiodically to affect forest structure. Whether biological or
environmental, they do not usually have equal impacts on all life-forms.

6.5.1 Wind

Typhoons (cyclones, hurricanes, willy-willies) are common in many parts
of the tropics and subtropics, especially the western side of warm ocean
basins at latitudes higher than about 10 °. Their impacts on life-form diversity can best be illustrated by the response of a single group, arborescent
palms. The broad crowns of dicotyledonous trees are architectural marvels
at intercepting light and occupying space, but they are almost equally
effective in intercepting hurricane-force winds, with the predictable outcome that they topple. Palms, in contrast, have two structural characteristics that enable them to survive wind storms. One is the elasticity and flexi-
bility of their stems (Wainwright 1976), which can bend to a remarkable
degree without either breaking or uprooting the palm. The second is that
the huge leaves of palms can be blown off, thus reducing the interception of

wind. These traits prompted Beard (1949) to suggest that the extensive, nearly monospecific *Prestoea* forests (palm brakes) on steep slopes of Caribbean islands were a reflection of the differential survival of this life-form to hurricanes, although poorly drained soils in these slopes exposed to the trade winds may be important as well (Frangi and Lugo 1985).

Winds less fierce than typhoons also influence tropical vegetation and the life-forms of which it is composed. Palm brakes (*Livistonia*) in Malaysia may reflect the combined drying effects of strong winds and shallow soil (Wyatt-Smith 1963), and Leigh et al. (1993) concluded that wind was a key factor in practically eliminating huge canopy trees from small, artificially created islands in Gatun Lake, Panama.

6.5.2 Fire

Despite their all-or-nothing investment in a single apical bud, palms are also remarkably resistant to fire. Again, the monocotyledonous anatomy plays a role. The vascular strands of palms are embedded throughout a matrix of parenchyma and fibers, which insulates the conducting tissues. The bud is often massive and sheathed in thick scales that similarly provide insulation. When fires sweep through forests dominated by a mix of dicotyledonous trees and palms, the dicots tend to be killed, leading to differential survival and eventual dominance by the palms (together with the hemiepiphytes they support). A remarkable example is the tens of thousands of hectares of monospecific stands of *Orbignya martiana* in southeastern Amazonia, a region blanketed by species-rich deciduous forest less than a century ago (Anderson et al. 1991).

Perhaps the most striking and widespread physiognomic change on tropical landscapes is the human-mediated conversion of forest to grass-land. Derived savannas, dominated by grasses such as *Imperata*, *Hyparrhenia*, *Panicum*, and *Saccharum*, occur throughout the tropics and are sustained by fire. Not surprisingly, such a dramatic change in life-form dominance results in major shifts in ecosystem functioning. The fires themselves yield emissions of gases important in climate change, and the retention of nutrients by shallow-rooted grasses is less effective than that of trees.

6.5.3 Animals

The changes induced in vegetation structure by animals range from whole-sale landscape conversion to minor shifts in composition and individual plant architecture. Among the most dramatic impacts on life-form diversity are those of large African herbivores: the conversion of savanna dominance from tall grass to short grass by hippopotamuses, for example. Perhaps the best known case is the conversion of woodlands to grasslands by elephants, which destroy trees in the process of consuming bark (e.g., Dublin et al. 1990). Not surprisingly, the dramatic changes in vegetation structure wrought by elephants cascade through the biota: tsetse flies breed in woodlands, and when the trees are killed the flies disappear; in the absence of tsetse flies, *Trypanosoma* (sleeping sickness) loses its vector, and the area becomes habitable by a number of mammals, including domesticated cattle, that would otherwise have been excluded by disease; and grazing potential attracts people, with all the changes their presence implies.

Less dramatic, perhaps, but equally important to ecosystem structure, is the Neotropical example studied by Leigh et al. (1993). They found that the absence of agoutis on small islands resulted in a proliferation of tree species whose seeds would otherwise have been consumed, and this contributed to reduced diversity of trees.

Sometimes the actions of an animal do not lead to replacement of one species or life-form by another, but to more subtle morphological changes in the targeted plant. Consider, for example, the effect of the marine isopod (*Sphaeroma*), whose larvae burrow into the exposed tips of mangrove (*Rhizophora*) prop roots. Death of the apex can lead to dichotomous branching, thus increasing the surface area of prop roots, which in turn serve as substrate for many sessile organisms, both plant and animal (Simberloff et al. 1978; but see Ellison and Farnsworth 1990). Or consider the impacts of white-faced monkeys (*Cebus*) on the architecture of *Gustavia* trees as a result of bud grazing. Consumption of the apex releases lateral buds, leading to a proliferation of new shoots, young leaves, and flowers and fruits (Oppenheimer and Lang 1969). In a dry area of southeastern Sri Lanka, elephant browsing deforms the crowns of the dominant trees, undoubtedly leading to shifts in the competitive balance among species (Mueller-Dombois 1972).

The impacts of one species, *Homo sapiens*, are so pervasive that they merit separate consideration, for there is scarcely a tropical forest in the world in which human activities have not left their mark on life-form diversity (Denslow and Padoch 1988; Goldhammer 1992). Nonindustrialized people selectively harvest large dicotyledonous trees for construction of bee hives and canoes (Kahumbu 1992) and vines as a raw material for basketry, tying, and construction. Palms tend to suffer disproportionately at the hands of humans because of singular features that lend themselves to a remarkable variety of uses: the large, fibrous leaves are used as thatch; the main axis of many climbing palms becomes rattan; the bud and starch reserves are sources of food; slats from the trunk are used in house construction; and fruits yield cooking oil, animal fodder, human food, stimulants, fertilizer, analytical-grade charcoal, organic chemicals, soap, ethanol, fuel, and many other products (Anderson et al. 1991).

Epiphytes, which reach their highest diversity and rates of endemism in montane belts, may be the life-form that is most threatened by anthropogenic disturbance. Nonvascular epiphytes in particular are less well known floristically than other groups, and there is little doubt that many species are extinguished before they are known to science. Because montane forests tend to be patchily distributed, destruction of relatively small areas can result in extinctions. Epiphyte populations are obviously affected by timber harvesting, even when it does not involve clearing of entire forest patches. More subtly, international demand for rare epiphytic orchids and bromeliads encourages the stripping of untold numbers of these plants from tropical forests. Furthermore, epiphytes, including lichens, are extremely sensitive to air pollution. Although levels of toxic emissions in most tropical regions are nowhere near as high as those that contributed to the decimation of the European bryoflora during the industrial revolution, the fertilization effects of nitrogen deposition due to agricultural practices and biomass burning in tropical regions may be causing insidious changes. The ability of bryophytic epiphytes to take up atmospheric sources of inorganic nitrogen could lead to long-term increases in nitrogen availability in the canopy via feedbacks to decomposition rates (Clark et al., unpubl. ms). In other nutrient-poor plant communities, increases in nitrogen supply have led to declines in species richness, a process possibly being replayed today in the canopies of tropical cloud forests.

6.5.4 Climate Change

The impacts of wind, fire, and wildlife on the life-form composition of tropical forests pale in comparison with the changes induced by swings of climate on a geologic time scale. Paleoecology has long been regarded by most as one of those interesting but esoteric endeavors of little direct relevance to society. Today, however, that perception has changed, as we become increasingly aware of the intimate relationships between human activity and climate.

According to P. Colinvaux (pers. comm.), three statements can now be made with reasonable assurance about the tropical lowlands of ice-age Earth: (1) temperatures were about 6 °C cooler than they are today, (2) atmospheric carbon dioxide concentrations were about two-thirds as high as today's, and (3) rainfall and seasonality were different – it was drier for longer in some places, and wetter in others. Paleoecologists recognize that vegetation types do not respond to climate change by remaining intact and slithering, amoeboid fashion, about the landscape. Rather, species drop in and out of communities, resulting in mixes that would seem incongruous in today's world: *Podocarpus* and *Magnolia* cohabiting with the palms and legumes of the modern lowland flora of the Neotropics, for example. There is no reason to think that the same sorts of recombinations that occurred with species did not also obtain at the level of life-forms, however hard it might be for us to visualize vine-festooned *Dacrydium*, or swards of grass sprinkled with terrestrial bromeliads, neither employing C3 carbon fixation. The shifts in life forms that most assuredly will accompany tropical climate change are bound to have profound effects on both primary productivity (through interactions between CO_2 and temperature effects by life-forms that employ different modes of carbon fixation, for example) and secondary productivity (because of the wide range in food supplying powers of different life forms – shrubs compared with treelets, for example). It is less certain that they will exert great effects on processes that are under more direct control of climatic and atmospheric conditions, such as soil development and fluxes of water and nutrients.

6.6 Life-Forms and Succession

Drawing on a 5-year study of post-slash-and-burn succession on a fertile soil in Costa Rica, we see that the vegetation undergoes important shifts in life-form composition, even in those early stages of stand development (Fig. 6.1, upper). A marked decrease in herbaceous vines, for example, was accompanied by increases in shrubs and trees. Perhaps the most dramatic change occurred with epiphytes, which underwent a dramatic surge in abundance during the 3rd year, when trees developed sufficient stature to support them (Fig. 6.1, lower). This is a clear case of facilitation, whereby colonization of an entire life-form was made possible by the development of structure provided by a second life-form.

Some life-forms, on the other hand, are major impediments to colonization and stand development. Vine tangles, like the grasses described earlier, sometimes hold sites against all potential invaders for decades. These life-form-poor disclimaxes may have very different functional traits than the richer forests that might otherwise occupy the same sites, but data to support or refute such an inference are scarce.

6.7 Implications of Loss of Life-Forms

Just as factors such as environment, episodic phenomena, and community development influence life-forms, the life-forms themselves influence the internal workings of ecosystems. We have illustrated how dominant life-forms influence ecosystem-wide resource availability and stand dynamics. Nevertheless, it is through their impacts on within-community functioning that the mix of life-forms is likely to have its greatest impacts. Some might argue that such within-community processes are bona fide components of ecosystem functioning, some might not; it is easy to imagine situations

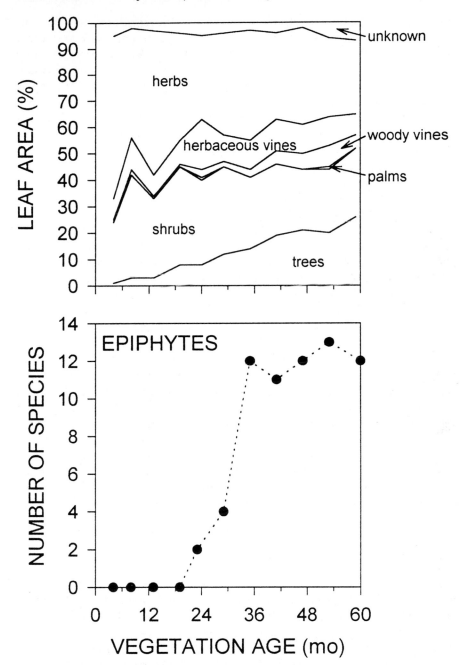

Fig 6.1. Changes in life-form composition during 5 years of post-slash-and-burn succession. The study took place near Turrialba, Costa Rica. Values are derived from species inventories and leaf-area assessments of six 16 x 16 m plots. *Upper* Relative dominance (expressed as proportional contribution of leaf area) by species comprising different life-forms; *lower* epiphyte colonization, facilitated by trees during the 3rd year

when either perspective would have merit. Semantic differences aside, the important point is that life forms do influence processes at many scales (just as different species, and different age classes and size classes within species, have distinct impacts on processes), and loss of a life-form is likely to precipitate changes in both the composition and functioning of the ecosystem of which it was a part.

Life-form diversity creates ecological opportunity, and this positive feedback culminates in the high biotic diversity we find in tropical forests. Rico-Gray (1993, and pers. comm.), for example, examined ant-plant associations in dry forest along the Gulf coast of Mexico. Associations between the 102 plant species and 30 ant species he studied sorted out by life form as follows: shrubs, 31 relationships with ants; trees, 26; herbs, 18; herbaceous vines, 10; woody vines, 9; parasites, 4; epiphytes, 2; cacti, 1; and fungi, 3. Although most of these relationships are nonspecific (one ant species fed on 72 plant species), it is likely that the diversity of relationships provides temporal continuity, and in some cases that continuity of food or defense is undoubtedly crucial to survival. Loss of shrubs, for example, might lead to local extinction of many ant species, whereas loss of cacti might be inconsequential.

Can we predict the consequences of loss of one life-form from a structurally diverse community, or is each situation so unique that we cannot generalize? As is the case with most ecological predictions, robust generalization may prove impossible due to the inherent complexity of interactions in biotic communities. Nonetheless, some life-forms fill certain roles uniquely (or nearly so; Table 6.2), and we can speculate about the consequences of their loss. As examples we choose two life-forms that reach peak abundance and diversity in tropical forests – woody vines and epiphytes.

Vines serve as trellises for arboreal mammals, such as rodents (Malcolm 1991; Langtimm 1992), monkeys (Emmons and Gentry 1983), and sloths (Montgomery and Sunquist 1978). They are also a webbing that binds tree crowns together, perhaps increasing canopy stability against windthrow, or perhaps generating larger openings when trees do fall. In addition to the benefits they impart on other members of the community, vines produce a tremendous amount of foliage which shades their host trees and slows their growth; furthermore, they sometimes add so much mass to a tree's crown that it breaks (Putz 1980). Other things being equal, a vine-free tropical forest might be expected to have fewer (or less mobile) arboreal mammals, faster tree growth, less tree breakage, and smaller gaps.

Table 6.2. Examples of linkages between plant life-forms and processes in tropical forests

Life-form	Role
Dicotyledonous trees, long-lived	1. Provide skeletal structure of entire forest 2. Dominate primary productivity and material flows 3. Influence off-site climate and hydrology 4. Provide shelter and roosts in hollow trunks
Dicotyledonous trees, short-lived	1. Reduce nutrient loss in early succession 2. Reduce likelihood of site takeover by vines and shrubs
Rosette trees (e.g., palms)	1. Channel rainwater toward stem 2. Capture and aggregate litter 3. Concentrate calcium 4. Roots bore through soil pans, creating channels that can be exploited by other plants 5. Root foraging emphasizes scale (rather than precision, sensu Campbell et al. 1991)
Understory trees	1. Scavenge sparse radiation of understory (and have low nitrogen demand) 2. Provide platforms (in humid microenvironment) for nitrogen-fixing epiphylls
Shrubs	1. Drive productivity of scansorial rodents and birds that feed on fleshy fruits 2. Retard nutrient loss in early succession
Giant-leaved herbs	1. Constitute large, homogeneous patches in otherwise heterogeneous understory 2. Foster secondary productivity through nectar and fruit production 3. Provide roosting sites for bats and building sites for carton nests of social insects
Vines	1. Provide trellises for movement of arboreal animals 2. Act as webbing that ties trees together 3. Buffer microclimatic changes by sealing forest edges
Graminoids	1. Constitute readily combustible dry-season fuel 2. Provide forage for grazers and food for seed-eating birds, rodents, ants, and fungi
Hemiepiphytes	1. Increase tree mortality rates 2. Provide slender vine trellises (aerial roots) in understory of closed canopy forest
Epiphytes	1. Augment leaf area (by colonizing opaque surfaces) 2. Slow nitrogen through-flow 3. Divert water from soil to atmosphere 4. Redistribute through-fall and stem flow 5. Provide unique habitats essential for reproduction of other species

Epiphytes, which scavenge water and nutrients out of the atmosphere, play important roles in stand hydraulics and nutrition. In forests where mist and rain come as pulses interspersed with dry spells, epiphytes act as capacitors, or storage devices. Much of the water that might otherwise plummet directly through the canopy as throughfall is captured by the epiphytes and eventually evaporates directly back to the atmosphere. In some forests this short-circuiting can be beneficial because it reduces water input to the soil, thus reducing leaching; in others, where water may be in short supply between rains, it can be detrimental. In continuously wet tropical forests, such as mist-shrouded cloud forests, bryophytic epiphytes are like permanently soaked sponges. They intercept vast quantities of saturated air, and the water condenses on the epiphytes. Nevertheless, because the epiphytes are already saturated, the water is not absorbed and stored, but drips off the epiphytes onto the forest floor. In one of the few stand-level life-form-removal experiments ever performed, Weaver (1972) found that removal of the epiphytes (dominated, in this case, by bryophytes) from a Puerto Rican dwarf cloud forest did not change the total amount of water reaching the forest floor, but it did produce an important change in its spatial distribution. In plots with epiphytes intact, water dribbled more or less evenly through the canopy, but in epiphyte-free plots, water was channeled down stems in torrents to the forest floor.

Inorganic nitrogen in cloud water, mist, and precipitation is rapidly taken up by epiphytic bryophytes; 70 to 80 % of the inorganic nitrogen in atmospheric deposition was retained by the forest canopy at one mid-elevation site (Clark et al., unpubl. ms). Much of this nitrogen, which is in a highly mobile form and arrives in pulses, is transformed into more recalcitrant forms and may persist for years as part of the mats of dead organic matter that line the limbs of large trees in montane cloud forest (Nadkarni and Matelson 1992). It appears likely, therefore, that nitrogen leaching losses in tropical montane forest ecosystems would be significantly greater in the absence of epiphytic bryophytes (Clark et al., unpubl. ms).

Tank-forming epiphytes are complex microcosms that support detritus-based food chains and are essential habitats for some stages of the life cycles of many organisms. Indeed, the study of these miniature communities, or phytotelmata, is an ecological subdiscipline in itself (Laessle 1961; Fish 1976). To cite one example, the males of a Central American terrestrial, viviparous frog, *Dendrobates pumilio*, transport young on their backs up trees to epiphytes, where the offspring leave the parent and develop into adults (Wells 1977). A second example concerns vectors of human disease. Because the larvae of some species of mosquitoes that spread yellow fever and malaria develop in them (Bates 1952; Pittendrigh 1948), the elimination of tank epiphytes was a major tactic in disease control earlier this century. Unfortunately, the ecological impacts of this landscape-level experiment were not evaluated.

Epiphytes, plus the water they store, add a tremendous mass to tree crowns and there can be no doubt that they promote limb breakage and tree fall. Epiphyte loads likely contribute to the rapid turnover of some tropical forests (e.g., Samper 1992), especially at mid-elevation where, due to orographic rainfall and the lifting condensation level, their abundance reaches its peak. Nevertheless, the contribution of epiphytes to tree fall is difficult to quantify because of the complexity of branching patterns, inter-specific differences among trees in their support of epiphytes, and spatial heterogeneity in the epiphyte flora.

Although epiphytes of many taxa form tanks of sorts in all tropical regions, none are as abundant or well-developed as those of the Bromeli-aceae, an essentially neotropical family. Thus, biogeography has yielded a giant, unreplicated experiment whose results have yet to be analyzed. Does the absence of bromeliad tanks in the African and Asian tropics, for exam-ple, manifest itself in measurable ways in inter-regional forest structure or functioning? Are certain functional groups of amphibians and insects less abundant because the microcosms required for completion of their life cycle are absent, or do they simply have different habitat requirements? Could the Dipterocarpaceae biomechanically support water-filled tank epiphytes, or would their dominance of Asian forests be reduced if brome-liads were present?

6.8 Conclusions

Our interpretation of ecosystem functioning as encompassing only fluxes and storages of energy and materials affords two advantages. First, these processes are definable and measurable (though seldom measured), and second, they are the processes whose influence is most likely to extend beyond the bounds of the system in question. From this admittedly narrow perspective, we conclude that it is the mix of life-forms, not the mix of species, that exerts major controls over ecosystem functioning.

Just as taxonomists base their science on phylogenetic relationships among the groups they study, ecologists should ground theirs on functional similarities. Unfortunately, our capacity to recognize architectural types far exceeds our capacity to measure functional differences among them. Nev-ertheless, categorization of life forms should be based on gross morphol-ogy, which in turn must be coupled to measures more directly linked to function – physiology, stature, and longevity; and to be useful in a particu-lar setting, a life-form categorization scheme must be tailored to the appropriate environment, in this case tropical forests.

The consequences of loss of life-form diversity can be predicted through experimentation, natural history observations, or analysis of landscape patterns. Rigorous determination of the relationships between life-form diversity and ecosystem functioning can come only from ecosystem-level experiments that include manipulation of life-form composition coupled with measurement of the appropriate response variables. Meanwhile, a number of natural "experiments" might yield qualitative insights into the consequences of life-form enrichment or impoverishment. These include biogeographical comparisons (e.g., between islands and mainlands, or between Neotropics and Paleotropics); analysis of patterns along gradients of climate and soil, both past and present; study of episodic and catastrophic events, such as typhoons, rain-forest fires, and imbalances in wildlife populations; and, perhaps most importantly, the disruptions of ecosystem structure and functioning caused by that most disruptive of species, *Homo sapiens*.

Acknowledgments. We thank Paul Colinvaux, who contributed everything paleo, plus several other nuggets, to this manuscript; Francis Putz, whose vast knowledge of vines, palms, roots, hemiepiphytes, and forest dynamics was tapped constantly by us; Victor Rico-Gray, who generously provided data on ant-plant relationships; Kenneth Clark, who shared his insights on epiphytes and their relationships to atmospheric chemistry; Ankila Hiremath, whose suggestions improved the clarity and accuracy of the manuscript; and F. Stuart Chapin III, who provided a very useful review. The views expressed herein reflect, in part, data and perspectives developed during work on National Science Foundation awards BSR-8506373, BSR-9000089, and BSR-9004281.

We dedicate this chapter to the memory of Alwyn H. Gentry, in recognition of his courageous devotion and unparalleled contributions to tropical botany, ecology, education, and conservation.

References

Anderson AB, May PH, Balick MJ (1991) The subsidy from nature: palm forests, peasantry, and development on an Amazon frontier. Columbia Univ Press, New York

Aubréville A (1938) La forêt coloniale: les forêts de l'Afrique occidentale française. Acad Sci Coloniales; Soc Editions Géographiques, Maritimes et Coloniales, Paris, Annales 9: 126-137

Bates M (1952) Where winter never comes: a study of man and nature in the Tropics. Scribner, New York

Beard JS (1949) The natural vegetation of the Windward and Leeward Islands. Oxford For Mem 21, 192 pp

Becker P, Castillo A (1990) Root architecture of shrubs and saplings in the understory of a tropical moist forest in Panamá Biotropica 22:242-249

Bormann FH, Berlyn G (eds) (1981) Age and growth rate of tropical trees: new directions for research. Yale Univ Press, New Haven, Connecticut

Böcher TW (1977) Convergence as an evolutionary process. Bot J Linn Soc 75:1-19

Box EO (1981) Macroclimate and plant forms: an introduction to predictive modeling in phytogeography. Dr W Junk, The Hague

Bullock SH (1985) Breeding systems in the flora of a tropical deciduous forest in Mexico. Biotropica 17:287-301

Chazdon RL (1985) Leaf display, canopy structure, and light interception of two understory palm species. Am J Bot 72:1493-1502

Chazdon RL (1991) Plant size and form in the understory palm genus *Geonoma*: are species variations on a theme? Am J Bot 78:680-694

Chazdon R, Fetcher N (1984) Photosynthetic light environments in a lowland tropical rain forest in Costa Rica. J Ecol 74:553-564

Clark DB, Clark DA (1991) The impact of physical damage on canopy tree regeneration in tropical rain forest. J Ecol 79:447-457

Connell JH (1978) Diversity in tropical rain forests and coral reefs. Science :1302-1310

Connell JH (1979) Tropical rain forests and coral reefs as open non-equilibrium systems. In: Anderson RM, Turner BD, Taylor RL (eds) Population dynamics. 20th Symp Br Ecol Soc, Blackwell, Oxford, pp 141-163

Connell JH, Slatyer RO (1977) Mechanisms of succession in natural communities and their role in community stability and organization. Am Nat 111:1119-1144

Cuatrecasas J (1958) Aspectos de la vegetación natural de Colombia. Rev Acad Cienc Nat Colombia 10:221-262

De Castro y Santos A (1980) Essai de classification des arbres selon leur capacité de réiteration. Biotropica 12:187-194

Denslow JS, Padoch C (eds) (1988) People of the tropical rain forest. Univ California Press, Berkeley, California

Dublin HT, Sinclair ARE, McGlade J (1990) Elephants and fire as causes of stable states in the Serengeti-Mara woodlands. J Anim Ecol 59:1147-1164

Ellenberg H, Mueller-Dombois D (1967) Tentative physiognomic-ecological classification of plant formations of the earth. Ber Geobot Inst Rübel 37:21-55

Ellison AM, Farnsworth EJ (1990) The ecology of Belizean mangrove-root fouling communities. I. Epibenthic fauna are barriers to isopod attack of red mangrove roots. J Exp Mar Biol Ecol 142:91-104

Emmons LH, Gentry AH (1983) Tropical forest structure and the distribution of gliding and prehensile-tailed vertebrates. Am Nat 121:513-524

Eriksson O, Bremer B (1991) Fruit characteristics, life forms, and species richness in the plant family Rubiaceae. Am Nat 138:751-761

Ewel JJ (1986) Designing agricultural ecosystems for the humid tropics. Annu Rev Ecol Syst 17:245-271

Fish D (1976) Structure and composition of the aquatic invertebrate community inhabiting epiphytic bromeliads in south Florida and the discovery of an insectivorous bromeliad. PhD Diss, Univ Florida, Gainesville

Fisher JB (1986) Branching patterns and angles in trees. In: Givnish TJ (ed) On the economy of plant form and function. Cambridge Univ Press, Cambridge, pp 493-524

Foster RB (1990) The floristic composition of the Río Manú floodplain forest. In: Gentry AH (ed) Four neotropical rainforests. Yale Univ Press, New Haven, Connecticut, pp 99-111

Fox JED (1968) Logging damage and the influence of climber cutting prior to logging in the lowland dipterocarp forest of Sabah. Malays For 1:326-347

Frangi JL, Lugo AE (1985) Ecosystem dynamics of a subtropical floodplain forest. Ecol Monogr 55:351-369

Gentry AH (1983) Dispersal ecology and diversity in neotropical forest communities. Sonderb Naturwiss Ver Hamburg 7:303-314

Gentry AH (1988) Changes in plant community diversity and floristic composition on environmental and geographical gradients. Ann Mo Bot Gard 75:1-34

Gentry AH (1991) The distribution and evolution of climbing plants. In: Putz FE, Mooney HA (eds) The biology of vines. Cambridge Univ Press, Cambridge, pp 3-49

Gentry AH, Dodson CH (1987a) Contribution of nontrees to species richness of a tropical rain forest. Biotropica 19:149-15

Gentry AH, Dodson CH (1987b) Diversity and biogeography of neotropical vascular epiphytes. Ann Mo Bot Gard 74:205-233

Goldhammer JG (ed) (1992) Tropical forests in transition: ecology of natural and anthropogenic disturbance processes. Birkhäuser, Basel Gradstein SR, Pócs T (1989) Bryophytes. In: Lieth H, Werger MJA (eds) Rain forest ecosystems. Elsevier, Amsterdam, pp 311-325

Grime JP, Hunt R (1975) Relative growth-rate: its range and adaptive significance in a local flora. J Ecol 63:393-422

Grubb PJ (1977) Control of forest growth and distribution on wet tropical mountains: with special reference to mineral nutrition. Annu Rev Ecol Syst 8:83-107

Grubb PJ, Stevens PF (1985) The forests of the Fatima basin and Mount Kerigomna, Papua New Guinea with a review of montane and subalpine rainforests in Papuasia. Aust Nat Univ, Canberra

Grubb PJ, Tanner EVJ (1976) The montane forests and soils of Jamaica: a reassessment. J Arnold Arbor 57:313-368

Grubb PJ, Lloyd JR, Pennington TD, Whitmore TC (1964) A comparison of montane and lowland rain forest in Ecuador. I. The forest structure, physiognomy, and floristics. J Ecol 51:567-601

Haggar JP, Ewel JJ (1994) Experiments on the ecological basis of sustainability: early findings on nitrogen and phosphorus, and root systems. Interciencia 19:347-351

Hall JB, Okali DUU (1978) Observer-bias in a floristic survey of complex tropical vegetation. J Ecol 66:241-249

Hallé F, Oldeman RAA, Tomlinson PB (1978) Tropical trees and forests: an architectural analysis. Springer, Berlin Heidelberg New York

Hartshorn GS, Poveda LJ (1983) Checklist of trees. In: Janzen DH (ed) Costa Rican natural history. Univ Chicago Press, Chicago, pp 158-183

Holbrook NM, Sinclair TM (1992) Water balance in the arborescent palm, *Sabal palmetto*. II. Transpiration and stem water storage. Plant Cell Environ 15:401-409

Hubbell SP (1979) Tree dispersion, abundance, and diversity in a tropical dry forest. Science 203:1299-1309

Hubbell SP, Foster RB (1983) Diversity of canopy trees in a neotropical forest and implications for conservation. In: Sutton SL, Whitmore TC, Chadwick AC (eds) Tropical rain forest: ecology and management. Blackwell, Oxford, pp 25-41

Kahn F (1986) Life forms of Amazonian palms in relation to forest structure and dynamics. Biotropica 18:214-218

Kahumbu P (1992) The sustainability of fig tree (*Ficus sycomorus*) harvesting for canoes in a Kenyan reserve. Master's Thesis, Univ Florida, Gainesville

Kohyama T (1987) Significance of architecture and allometry in saplings. Funct Ecol 1:399-404

Laessle AM (1961) A microliminological study of Jamaican bromeliads. Ecology 42:499-517

Langtimm CA (1992) Specialization for vertical habitats within a cloud forest community of mice. PhD Diss, Univ Florida, Gainesville

Leigh EG Jr (1975) Structure and climate in tropical rain forest. Annu Rev Ecol Syst 6:67-86

Leigh EG Jr, Wright SJ, Herre EA, Putz FE (1993) The decline of tree diversity on newly isolated tropical islands: a test of a null hypothesis and some implications. Evol Ecol 7:76-102

MacArthur RH, Wilson EO (1967) The theory of island biogeography. Princeton University Press, Princeton, New Jersey

Malcolm JR (1991) The small mammals of Amazonian forest fragments. PhD Diss, University of Florida, Gainesville

Matelson TJ, Nadkarni NM, Longino JT (1993) Longevity of fallen epiphytes in a neotropical montane forest. Ecology 74:265-269

May RH (1975) Stability in ecosystems: some comments. In: Van Dobben WH, Lowe-McConnell RH (eds) Unifying concepts in ecology. Dr W Junk, The Hague, pp 161-168

Montgomery GG, Sunquist ME (1978) Habitat selection and use by two-toed and three-toed sloths. In: Montgomery GG (ed) The ecology of arboreal folivores. Smithsonian Inst Press, Washington DC, pp 329-359

Moore HE (1973) Palms in the tropical forest ecosystems of Africa and South America. In: Meggers B, Ayensu E, Duckworth W (eds) Tropical forest ecosystems in Africa and South America: a comparative review. Smithsonian Inst Press, Washington DC, pp 63-68

Mueller-Dombois D (1972) Crown distortion and elephant distribution in the woody vegetation of Ruhuna National Park, Ceylon. Ecology 53:208-226

Mueller-Dombois D, Ellenberg H (1974) Aims and methods of vegetation ecology. Wiley, New York

Mueller-Dombois D, Cooray RG, Maka JE (1981) Forest layers and quantitative composition of the plant community. In: Mueller-Dombois D, Bridges KW, Carson HL (eds) Island ecosystems: biological organization in selected Hawaiian communities. Hutchinson Ross, Stroudsburg, Pennsylvania, pp 231-246

Myers RL (1990) Palm swamps. In: Lugo AE, Brinsen M, Brown S (eds) Forested wetlands, vol 15. Ecosystems of the world. Elsevier, New York, pp 267-286

Nadkarni NM, Matelson TJ (1992) Biomass and nutrient dynamics of epiphytic litterfall in a neotropical cloud forest, Monteverde, Costa Rica. Biotropica 24:24-30

Oppenheimer JR, Lang GE (1969) *Cebus* monkeys: effect on branching of *Gustavia* trees. Science 165:187-188

Pearl R (1928) The rate of living, being an account of some experimental studies on the biology of life duration. Knopf, New York

Pittendrigh CS (1948) The bromeliad-Anopheles-malaria complex in Trinidad. I. The bromeliad flora. Evolution 2:58-89

Proctor J, Anderson JM, Chai P, Vallack HW (1983) Ecological studies in four contrasting lowland rain forests in Gunung Mulu National Park, Sarawak. I. Forest environment, structure, and floristics. J Ecol 71:237-260

Putz FE (1980) Lianas vs. trees. Biotropica 12:224-225

Putz FE (1983) Liana biomass and leaf area of a "Tierra Firme" forest in the Rio Negro basin, Venezuela. Biotropica 15:185-189

Putz FE (1985) Woody vines and forest management in Malaysia. Comm For Rev 64:359-365

Putz FE,Chai P (1987) Ecological studies of lianas in Lambir National Park, Sarawak, Malaysia. J Ecol 75:523-531

Ramírez N (1993) Producción y costo de frutos y semillas entre formas de vida. Biotropica 25:46-60

Raunkiaer C (1934) The life forms of plants and statistical plant geography. Clarendon Press, Oxford

Rico-Gray V (1993) Use of plant-derived food resources by ants in the dry tropical lowlands of coastal Veracruz, Mexico. Biotropica 25:301-315

Rockwood LL (1985) Seed weight as a function of life form, elevation, and life zone in neotropical forests. Biotropica 17:32-39

Rose MR (1991) Evolutionary biology of aging. Oxford Univ Press, New York

Samper C (1992) Natural disturbance and plant establishment in an Andean cloud forest. PhD Diss, Harvard Univ, Cambridge

Shreve F (1936) The transition from desert to chaparral in Baja California. Madroño 3:257-320

Simberloff D, Brown BJ, Lowrie S (1978) Isopod and insect root borers may benefit Florida mangroves. Science 201:630-632

Solbrig OT (1993) Plant traits and adaptive strategies: their role in ecosystem function. In: Schulze E-D, Mooney HA (eds) Biodiversity and ecosystem function. Springer, Berlin Heidelberg New York, pp 97-116

Stevens GC, Perkins AL (1992) The branching habits and life history of woody plants. Am Nat 139:267-275

Stocker GC (1988) Tree species diversity in rainforests - establishment and maintenance. In: Kitching R (ed) The ecology of Australia's wet tropics, vol 15. Proc Ecol Soc Australia. Surrey Beatty and Sons, Chipping Norton, Australia, pp 39-47

Vitousek P (1985) Community turnover and ecosystem nutrient dynamics. In: Pickett STA, White PS (eds) The ecology of natural disturbance and patch dynamics. Academic Press, San Diego, pp 325-333

Vogel S (1972) Exercise 12. In: Vogel S, Ewel KC (eds) A model menagerie: laboratory studies about living systems. Addison-Wesley, Reading, Massachusetts

Wainwright SA (1976) Mechanical design in organisms. Wiley, New York

Warming E (1909) Oecology of plants. Clarendon Press, Oxford

Weaver PL (1972) Cloud moisture interception in the Luquillo mountains of Puerto Rico. Carib J Sci 12:129-144

Weinstein DA, Shugart HA (1983) Ecological modeling of landscape dynamics. In: Mooney HA, Godron M (eds) Disturbance and ecosystems. Springer, Berlin Heidelberg New York, pp 29-46

Wells K D (1977) The courtship of frogs. In: Taylor DH, Guttman SI (eds) The reproductive biology of amphibians. Plenum, New York, pp 233-262

Whitmore TC, Peralta R, Brown K (1985) Total species count in a Costa Rican tropical rain forest. J Trop Ecol 1:375-378

Whittaker RH and Niering WA (1965) Vegetation of the Santa Catalina mountains, Arizona: a gradient analysis of the south slope. Ecology 46:429-452

Wyatt-Smith J (1963) Manual of Malay silviculture for inland forest (2 vols). Malayan For Rec No 23

7. Functional Group Diversity and Responses to Disturbance

Julie S. Denslow[1]

7.1 Introduction

In tropical moist forests, disturbance may enhance the coexistence of species through reduction of competition by dominants (Connell 1978; Huston 1979) and creation of establishment sites for seedlings (Brokaw 1985; Denslow 1980a, 1987). However, intense, repeated, and extensive human disturbance causes high rates of species extinctions (Wilson 1988). The debate on how best to manage tropical forests commercially shows that the relations between disturbance patterns, species diversity, and ecosystem processes are not well understood (Whitmore and Sayer 1992). In particular, the effects of species richness on disturbance regimes and successional patterns are more poorly understood than the effects of disturbance on community composition. This chapter examines the effects of species and groups of species on both disturbance regimes and the successional processes they trigger.

Tropical rain forests are diverse in species and in their spatial and temporal distribution. They also include many kinds of species. Plants and animals have been variously clustered into nontaxonomic groups based on adaptive syndromes (reviewed by Solbrig 1993), such as structure (Halle et al.1978; Ewel and Bigelow, Chap.6, this Vol.), position of perennating organs (Raunkiaer 1934); physiological characteristics (Bazzaz and Pickett 1980), foraging mode (Root 1967), and life-history strategies (Grime 1977, Huston and Smith 1987). All such groups are basically arbitrary but they may be useful for specific purposes.

Here I suggest that functional groups may help us understand the effects of plant types on forest dynamics and regeneration patterns. This chapter proposes a classification of functional groups designed to further our understanding of how species assemblages affect responses to disturbance, shows how such a classification system might be used, and reaches some conclusions about why functional group diversity is important in tropical forest dynamics.

[1] Department of Botany, Louisiana State University, Baton Rouge, Louisiana 70803, USA

Ecological Studies, Vol. 122
Orians, Dirzo and Cushman (eds) Biodiversity and Ecosystem Processes in Tropical Forests
© Springer-Verlag Berlin Heidelberg 1996

7.2 Functional Groups Affecting Tropical Forest Dynamics

The basis for recognizing functional groups with respect to their effects on stand dynamics is that species that share morphological, chemical, structural or life history characteristics are likely to affect stand dynamics in a similar way. For example, plant species with similarities in canopy structure, foliage characteristics, or maximum size are likely to exert similar effects on patterns of light availability, on soil characteristics, litter depth, and litter quality, or on disturbance patterns, respectively. These species may also similarly affect the rate and direction of succession following both natural and anthropogenic disturbances. Table 7.1 is one such classification, which is fairly complete for plants but is illustrative rather than exclusive or exhaustive for animals. Groups are described for tropical rain forests in which the major environmental limitation to plant growth is probably light availability. Different functional groups may be more appropriate for other ecosystems, such as seasonally deciduous tropical forest, in which other limitations to plant growth are important.

7.2.1 Pioneer Herbs and Shrubs

These plants are usually associated with human disturbance and are neither diverse nor abundant where human intervention in rain forest is slight. For example, the majority of species listed in this group for Barro Colorado Island, Panama, are most abundant in the laboratory clearing or on the lake margin rather than within the forest (Croat 1978). These species follow logging roads into the forest, are high-light-demanding, prolific reproducers that set large quantities of well-dispersed seed and often spread vegetatively as well. Consequently, they tolerate repeated clearing, fire, and grazing better than trees and dominate a frequently disturbed site within a few years (Kowal 1966; Kellman and Adams 1970). Their growth is suppressed when they are overtopped by taller plants.

7.2.2 Large-Leaved Understory Herbs and Shrubs

This group consists of low stature, shade-tolerant species that produce a deep shade at the ground (Denslow et al 1991). They depress the growth and survival of seedlings by increasing their susceptibility to insects and pathogens (Augspurger 1984).

Stands in which large-leaved understory plants are abundant often have low densities or patchy distributions of tree seedlings (Denslow et al 1991). Because this pool of seedlings rather than a soil bank of dormant seeds dominates regeneration of tropical forests(Uhl et al 1988), factors affecting the abundance, distribution, and diversity of the seedling pool influence stand regeneration patterns following both treefalls and selective logging. Within this group are shrubs and large herbs in the families Cycadaceae, Cyclanthaceae, Palmae, Musaceae, Zingiberaceae, Heliconiaceae, Marantaceae, and Araceae.

Table 7.1. Functional groups affecting rain forest dynamics in the tropics. Organisms are grouped according to their effects on disturbance regimes and forest regeneration processes

I. Plants

Group	Characteristics	Effects on forest dynamics
A. Herbs and Shrubs		
1. Pioneer	Incl. grasses and ferns; high light requirements, low nutrient tolerance	Colonization of large gaps, landslides, skid tracks, old fields, pastures; can suppress tree seedlings
2. Understory		
a. Large-leaved	Shade-tolerant herbs and shrubs; often palm-like	Heavy shade suppresses seedlings in understory
b. Small-leaved	Shade-tolerant, multi-stemmed herbs, shrubs	Competition with seedling, and saplings
B. Treelets	Small trees; generally not long-lived	
1. Pioneer	High light-demanding, short lived, copious small seeds	Rapid growth in large canopy openings and clearings Suppress growth of pioneer herbs and shrubs; affect micro-environment of early succession
2. Understory	Shade tolerants, sub-canopy trees; resprouters	Compete with saplings of canopy trees
C. Canopy and emergent trees	Big trees, long-lived, shade-tolerant or light-demanding	
1. Legumes	Often dominant plant family, high N litter	Increase decomposition rate and availability of nutrients
2. Palms	Voluminous, high-fiber litter; dense shade	Decrease litter decomposition rates; colonize landslides
3. Emergents	Very large trees with crowns projecting above canopy; long-lived but may be fast-growing	Create large treefall gaps and patches of high light
D. Lianas and vines		
1. Lianas	Large woody vines; long-lived, high-light response; often connect many tree crowns	May increase treefall gaps; high litter production; rapid growth in large gaps; compete with saplings
2. Vines	Herbaceous, high-light- or shade-tolerant	Grow rapidly in gaps, landslides, skid tracks

Table 7.1. (continued)

Group	Characteristics	Effects on forest dynamics
E. Epiphytes and hemiepiphytes		
1.Nonparasitic herbs	Herbaceous, canopy habitat	Sequester nutrients; weight may increase branch falls
2.Parasitic and hemi-epiphytic trees and shrubs	Woody; sometimes stranglers or vascular parasites	May contribute to death of canopy trees

II. Fungi, microbes, and animals

Group	Taxonomic composition	Effects on forest dynamics
A. Seed dispersers	Fruit-eating birds and mammals	Disperse seeds within forests and across landscapes
B. Pathogens and herbivores	Parasitic fungi, herbivorous insects, and some vertebrates (e.g., peccaries pacas, possums)	Affect vigor and mortality of plants of all sizes, but especially establishing seedlings in the understory
C. Soil processers		
1.Decomposers	Fungi, microbes, soil mesofauna	Increase nutrient supply rates; affect soil structure, incl. nutrients, moisture, and oxygen availability
2.Mycorrhizal fungi	Symbiotic associations with plants; endotrophic or ectotrophic	Increase availability of P and perhaps other nutrients; decompose organic matter (ectotrophic); increase seedling survival.
3.Soil churners	Animals foraging in litter and top soil (incl. ground-feeding birds, peccaries, leaf-cutter ants)	Remove litter, aerate soils, may create establishment sites for small seeded species; also kill seedlings

7.2.3 Small-Leaved Understory Herbs and Shrubs

These plants are also competitors for light with tree saplings. Where they are abundant, the density and diversity of tree saplings are low (Denslow and Chaverri, unpubl). Low carbon allocation to wood, horizontal branching patterns, and clonal growth, common in this group, also promote rapid occupancy of space in treefall gaps (Denslow et al. 1990 and unpubl). These characteristics may also facilitate their persistence in highly shaded understory environments where they may quickly grow into new canopy openings.

7.2.4 Pioneer Trees

This group consists of both short- and long-lived species with low shade tolerance and rapid establishment capabilities. In old-growth forests, trees in this group occur at low densities, confined to large canopy openings or permanently disturbed areas such as stream banks (Hubbell and Foster 1986a). In anthropogenic successions, they may form an even-aged, homogeneous tree canopy which affects the quality and distribution of light in the understory (Saldarriaga and Luxmoore 1991), the pattern of biomass and leaf-area accumulation, and the characteristics of soil organic matter and nutrient turnover rates (Weaver et al. 1987; Lugo 1992).

These species produce crops of copious, small, well-dispersed seeds over long fruiting seasons (Howe and Smallwood 1982). Although these seeds may have extended dormancy capacities, soil stocks are nevertheless rapidly depleted by seed predation (Alvarez-Buylla and Garcia-Barrios 1991) and germination following repeated clearing (Uhl et al. 1982; Young et al. 1987). The early dominance of this group in fallows following pasture or long-term agricultural use thus may depend critically on the proximity of a seed source, the abundance and behavior of vertebrate seed dispersers, and the ability of seedlings to establish in competition with grasses (Uhl 1988).

7.2.5 Understory Treelets

These are small trees whose crowns do not reach the main forest canopy. Although the growth of many trees is plastic in response to variation in light, growth rates of these species are low across a wide range of light conditions. They contribute to the dense shade in rain forest interiors and are competitors for light with saplings of canopy trees. Some resprout readily following stem breakage. Lieberman et al. (1985) suggest that these species turn over more rapidly than other groups of trees because of their susceptibility to damage from falling limbs and canopy trees (Clark 1991).

7.2.6 Emergent and Canopy Trees

Gap sizes and microclimatic conditions within gaps are largely a function of the sizes of gap-forming trees (Denslow and Hartshorn 1994). Old-growth, lowland, tropical rain forests are characterized by a suite of species whose crowns project and spread above the main canopy of the forest. Although large trees occur at relatively low densities (Clark and Clark 1992), their crowns may cover a relatively high proportion of the area. Gaps created by the fall of these giants are correspondingly large and bright. The abundance, diversity, and crown characteristics of tree species capable of reaching these very large sizes may thus affect canopy and stand turnover rates as well as the heterogeneity of light environments within the forest.

For example, in equatorial Africa, Hart et al (1989) found lower rates of treefall in forests dominated by narrow-crowned *Gilbertiodendron dewevrei* than in adjacent diverse forest on similar sites. Rankin (1978) suggested that forests dominated by *Mora excelsa* in Trinidad were less affected by hurricanes than were adjacent heterogeneous forest on similar sites. She hypothesized that the homogeneous *Mora* canopy was more resistant to wind damage than the naturally dissected canopy of the diverse forest. It is likely that other monodominant forests fit this pattern as well (Connell and Lowman 1989). Malaysian forests dominated by narrow-crowned trees in the Dipterocarpacae (Whitmore 1984) may be characterized by smaller gap sizes than are forests dominated by the Leguminosae (e.g., South America, Gentry 1982), which produce comparatively broader crowns both as emergents and as main canopy species. Similarly, the topographically smooth canopies of secondary forests (Brokaw 1982; Saldarriaga et al. 1988; Saldarriaga and Luxmoore 1991) and the small crowns typical of hurricane forests (Scatena and Lugo, unpubl.) may be less susceptible to windthrow and therefore turn over at lower rates than do more heterogeneous canopies.

7.2.7 Canopy Palms

The singular growth form and large leaf size of canopy palms contribute to their effect on disturbance processes. Where canopy palms are common, much tree death may occur with little effective canopy opening because palm crowns are small. Their voluminous, slowly decomposing litter contributes to seedling mortality and suppresses germination (Clark 1991). Palms often lose their leaves in high winds, lowering their susceptibility to windthrow and breakage (Walker 1991). As a consequence, they are able to replace their crowns quickly following hurricanes. The ability of palms to establish on unstable, nutrient-poor soils or other environmentally stressful situations is evidenced by their abundance on landslides, waterlogged soils, and disturbed forests.

7.2.8 Canopy Legumes

Canopy trees, emergents, and lianas produce the majority of litter in a tropical forest (Sanford unpubl.). Because litter quality strongly affects decomposition rates, high-nitrogen, legume litter decomposes more rapidly than litter high in lignin, phenols, and celluloses, such as that produced by palms or other high-fiber species. One of the consequences of legume dominance in Neotropical forests (Gentry 1982) may be low N immobilization by microorganisms, rapid nitrogen turnover rates, and little nitrogen limitation to plant growth in these ecosystems (Vitousek 1984; Blair et al. 1990, Bloomfield et al. 1993). Supply rates of other nutrients may be high as well. Rapid vegetation growth following canopy opening in many tropical moist forests reflects relatively high nutrient supply rates (Weaver et al. 1987; Brown and Lugo 1990).

7.2.9 Vines and Lianas

Lianas bind together tree crowns, increasing the likelihood that a treefall will bring down neighboring trees as well (Putz 1984). Vines and lianas are an important component of canopy closure in large, bright gaps where they may slow the establishment of a tree canopy (Putz and Mooney 1991) and create substantial management problems for tropical silviculturists.

7.2.10 Epiphytes

A diverse and well-developed epiphyte community is characteristic of old-growth tropical moist forests. Where epiphyte biomass (and water-holding capacity) is high (as in tropical montane forests), limb breakage and failure rates of large trees may also be high (Strong 1977). In contrast, tank epiphytes, such as water-holding bromeliads, are rare in Paleotropical forests. Parasitic shrubs, stranglers, and other hemiepiphytes may contribute to rates of tree death where they are abundant.

7.2.11 Seed Dispersers

In tropical moist forest, the majority of plant species dispersed by animals are carried by birds and bats (Howe and Smallwood 1982), although in the African and Asian tropics large mammals (especially elephants and primates) may play a substantial role as well (Whitmore 1984). These animals provide the mobile links (sensu Gilbert 1980) across the landscape, affecting the abundance and diversity of the propagule pool, and potentially affecting successional patterns as well. In the Asian tropics, many tree species, such as those in the family Dipterocarpaceae, an important family of canopy trees, are wind dispersed. Accumulating evidence suggests that animals are more efficient dispersers than wind (Howe and Smallwood 1982) and most pioneer species are animal- rather than wind-dispersed.

Morphological and behavioral variation among dispersers suggests that this group might be usefully subdivided. For example, gape size of birds affects the sizes of seeds that can be ingested and transported (Moermond and Denslow 1985; Wheelwright 1985), and canopy-dwelling birds move over longer distances and are more likely to forage and deposit seeds in open areas than are understory species (Stiles 1988; Loiselle and Blake 1991).

7.2.12 Herbivorous Insects and Pathogens

In the context of forest dynamics, herbivores and pathogens appear to be an important source of mortality for shaded seedlings (Augspurger 1984;

Augspurger et al. 1988) and subcanopy trees, and may hasten the senescence of early succession dominants. High stand turnover rates reported from tropical forests (Rankin-de-Merona et al. 1990) are due in large part to high mortality rates within the small diameter size classes (Clark and Clark 1992). Pathogen stress and insect herbivory probably contribute to the mortality of these trees, but there have been few studies of the effects of pathogens on stand dynamics in rain forest communities.

7.2.13 Decomposers

Microbes, fungi, and soil mesofauna affect rates of fragmentation and decomposition of organic matter and mineralization rates of nutrients. Their activity affects rates of nutrient supply, soil structure, site suitability for seed germination, and rates of plant growth (Jordan 1985). For example, the density of soil macropores created in part by foraging of earthworms may affect soil drainage and nutrient leaching (Sollins and Radulovich 1988).

7.2.14 Mycorrhizal Fungi

These species are symbiotic on the roots of most terrestrial vascular plants in tropical forests (Lodge et al., Chap.5, this Vol.). Ectotrophic, but not endotrophic, mycorrhizal fungi directly contribute to the breakdown of organic matter and uptake of nutrients by the host plants. Endotrophic mycorrhizae facilitate the uptake of phosphorus and perhaps other nutrients primarily through efficient exploration of the soil volume by the fungal mycelium (Read 1991). Seedlings grow slowly in the absence of a mycorrhizal association (Janos 1980) and low infection rates of seedlings in pastures or landslides may limit colonization rates of these sites by woody species (Janos 1980). Connell and Lowman (1989) have proposed that host specificity among ectomycorrhizal fungi may limit the invasibility of sites to new species and promote high levels of dominance in some tropical rain forests.

7.2.15 Soil-Churning Animals

The availability of exposed mineral soil is important to the establishment of many small-seeded species (Brandani et al. 1988, Ellison et al. 1993). In addition to uprooted trees, patches of disturbed soil are created by leafcutter ants, ground-dwelling rodents, and foraging peccaries, pigs, tapirs, deer, and bovines. Mammals and birds that forage in the leaf litter and surface soils may also kill or damage seedlings.

7.3 Functional Groups and Natural Disturbance Processes in Tropical Moist Forests

In tropical rain forests, disturbance events occur at varying spatial scales including soil churning by leaf-cutter ants (at a spatial scale of $1-100$ m^2), foraging peccaries and other vertebrates ($1-1000$ m^2), landslides ($100-10\,000$ m^2), and river meanders ($1000-10\,000$ m^2), and canopy opening due to branchfalls ($10-100$ m^2), treefalls ($100-1000$ m^2), and tropical storms ($1\,000\,000$ m^2). At larger scales and longer intervals, drought associated with El Niño/Southern Oscillation (ENSO) events or climate change in geological time may affect regional disturbance patterns (Saldarriaga and West 1986, Goldhammer and Seibert 1989). In old-growth or primary rain forests, the disturbance regime is a product of the interaction of abiotic (climate, topography, geology, soil characteristics) and biotic (functional group) processes. Some functional groups (e.g., emergent trees, lianas, epiphytes) may directly affect disturbance patterns as well as modify the effects of abiotic influences. The disturbance regimes of anthropogenic landscapes may be similarly described, although these are likely to be more variable than natural regimes and dominated by human social and economic processes.

Although many people believe that undisturbed tropical moist forests are dominated by trees of great age, size, and permanence, these forests are among the most dynamic of tree communities. Canopy turnover times (due to gap formation) are commonly around 100 years in old growth forest (Denslow and Hartshorn 1994). Tree (or stand) turnover times are much less. Rankin-de-Merona et al. (1990) estimate stand half-lives of about 40 years for many tropical forests. For small plants especially, growth and survival are strongly dependent on the high light environments in canopy openings formed by the fall of one or a few trees and sometimes on the presence of exposed mineral soil due to root-throw or animal disturbance (Ellison et al. 1993). Accumulating numbers of studies attest to the generality of this pattern across geographic ranges, community types, and species within the tropical moist forest biome (reviewed in Brokaw 1985; Denslow 1987). Moreover, frequent treefalls probably contribute to the maintenance of species richness through associated increases in both stem and species densities. High tree densities are often a product of high stand turnover rates. Sampling effects result in a higher number of species per unit area where densities are high than where they are low (Denslow, 1995).

In most tropical rain forests large-scale disturbances are rare. Large canopy openings (>500 m^2) originate in the fall of emergents; in the subsequent enlargement of smaller gaps due to effects of lianas or wind turbulence (Young and Hubbell 1991); and from cyclonic storms that fell large patches of trees. In most tropical forests, these large openings represent a small proportion of the gaps, and a somewhat larger proportion of the total area in gaps at any one time (e.g., Sanford et al. 1986). Some tropical moist forests, however, by virtue of their geographic location or topographic position, are frequently subjected to large-scale disturbances: in topographically dissected terrain landslides may be common (Garwood et al 1979; Guariguata 1990). Along major rivers, channel meanders create frequent open sites on which pioneer trees establish (Salo et al. 1986; Foster 1990). At a larger scale, forests in the tracks of tropical storms are devastated by landscape-scale destruction (Whitmore 1989, Walker 1991). Forests subject to frequent hurricanes are often dominated by high-light demanding species (Whitmore 1989; Denslow 1980b) and sometimes by single species (Connell and Lowman 1989).

Charcoal is commonly present in tropical rain forest soils (Sanford et al. 1985, Saldarriaga and West 1986; Goldhammer and Seibert 1989); however, the implications of these data are unclear because we do not know whether the charcoal was produced by small fires following lightning strikes, widespread fires in the landscape, or burning of swidden fields associated with traditional agricultural practices. In tropical dry and moist forests, storm or logging damage may be followed by fire (Goldhammer and Seibert 1989; Whigham et al. 1991). In some parts of the tropical Pacific, droughts associated with ENSO events may increase the probability and extent of forest fires (Goldhammer and Seibert 1989). However, fire is rare in undisturbed tropical moist forests today (e.g., Uhl and Kauffman 1990).

Few tropical rain forests appear to be structured by stand or forest-scale diebacks described for boreal and other coniferous forests (Holling et al. 1995). In boreal ecosystems, the structural and physiological attributes of old-growth forests (including accumulation of biomass, nutrient limitations, and insect or pathogen stress) increase the probability and magnitude of devastation at the scale of hectares and square kilometers of forest. In tropical forests, the probabilities of fires and insect outbreaks do not seem to be affected by forest structure or age. Because tropical forest trees often occur at low densities or in patchy distributions (Hubbell and Foster 1986b), pathogens or defoliating insects have localized rather than stand-wide effects (Janzen 1970). In some cases, pathogens and herbivores may reduce densities of dominants (Condit et al. 1992).

The outstanding exception to these generalizations is the 'ohi'a (*Metrosideros polymorpha*) forest dieback in Hawaii. 'Ohi'a dominates the canopy of montane rain forests under many different edaphic conditions. Massive diebacks occurring during the 1950s and 1970s have not been conclusively attributed to any single factor. Recent reviews suggest that stand-wide mortality is due to the senescence of cohorts under stress, but that a variety of factors might produce the stress, including pathogens, insects, ground fires, or nutrient limitation (Mueller-Dombois 1983; Stemmermann 1983). The invasion of exotic tree species able to establish under the open crowns of the 'ohi'a is increasing the species diversity of these forests and profoundly altering forest successional patterns (Vitousek et al. 1987; Vitousek and Walker 1989).

Physiological and environmental processes affecting the rates and patterns of plant growth in tropical treefall gaps have been reviewed elsewhere (Brokaw 1985; Denslow 1987; Denslow and Hartshorn 1994). Two conclusions from those reviews bear reiteration: first, shade-tolerant species of tropical moist forests are able to grow in a wide range of light environments, whereas survival of high-light-demanding species is poor in shade. Although growth rates of shade tolerants are often reduced in full sunlight, these species are able to establish and grow in extensive canopy openings such as those caused by hurricanes (Walker 1991; Yih et al. 1991). One year after treefall, more species are present in large, bright gaps at La Selva because seedlings of both high-light-demanding and shade-tolerant species are able to establish there. Only shade-tolerant species are able to persist for long in small gaps (Denslow unpubl.).

Second, for the most part, canopy openings created by natural disturbances (including large storms) are strongly buffered by remnants of the previous community and by the surrounding forest. Light, humidity, wind, and temperature are moderated by the adjacent tall forest, and remnants of the previous vegetation such as soil organic matter, seed stocks, seedlings, shrubs, and saplings insure that soil processes and forest community composition remain little modified (Jordan 1985; Vitousek and Denslow 1986; Uhl et al. 1988). As a result many species and functional groups present before the disturbance reestablish, and secondary succession following treefall is characterized primarily by changes in forest structure (leaf area index, biomass, density, height) rather than composition. Many of the functional groups described in Section 7.2 affect patterns of growth following canopy-opening disturbances. For example, understory shrubs and herbs may affect composition and abundance of tree seedlings present at treefall, and groups affecting nutrient supply (mycorrhizae, decomposers, legumes) may influence rates of canopy closure or gap filling.

7.4 Anthropogenic Disturbances to Tropical Forests

Anthropogenic disturbances likewise involve varying degrees and combinations of canopy opening and soil degradation (Whitmore and Sayer 1992). The salient difference between most natural and anthropogenic disturbances is the prolonged duration of human intervention. In general, treefalls, landslides, and even tropical storms are discrete events, although canopy openings may sometimes continue to enlarge (Young and Hubbell 1991) and slip surfaces may remain unstable for many years. Nevertheless, successional processes generally proceed with few interruptions by new disturbances.

In contrast, human exploitation of forest land is often characterized by repeated and prolonged intervention (Power and Flecker, Chap. 9, this Vol.). Swidden fields in the Neotropics are commonly farmed for 3-5 years followed by another 20 years or more in which the fallow is managed for fruit and fiber production (Redford and Padoch 1992). During the active phase, competing plants are repeatedly cleared by machete. Plantations and pastures are grazed and frequently cleaned or burned to control natural regrowth. Selectively logged forests are often reentered to extract more timber or cleared altogether by farmers or ranchers who follow the loggers (Whitmore and Sayer 1992). Land tenure and forestry laws coupled with low commercial value of forest products often result in little incentive for landowners to allow natural successional processes to reestablish forest on land that is no longer agriculturally productive.

When the soil is repeatedly exposed, the impact of rain, leaching of mobile ions, and the oxidation of organic matter result in soil compaction and nutrient depletion (Nye and Greenland 1961; Ewel 1983; Jordan 1985). Light levels are often much higher following anthropogenic disturbance than natural treefalls because canopy openings are larger and few remnant trees remain.

Tropical successions reflect site conditions, disturbance history, and propagule availability. As in temperate mesic forests, changes in dominance attributed to species replacements are often due to differential growth rates of species establishing following canopy opening (Drury and Nisbet 1973; Pickett 1976; Connell and Slatyer 1977). The composition of the seed pool in the early stages of regrowth potentially affects both vegetation patterns and ecosystem processes.

During early succession, density of stems >1 cm dbh first increases, then decreases as the stand increases in height and basal area (Crow 1971; Christensen and Peet 1981; Saldarriaga et al. 1988; Brown and Lugo 1990). Tropical secondary forests accumulate woody species rapidly during the first 80 years (Knight 1975; Brown and Lugo 1990), but, unlike temperate successions, diversity continues to increase during later phases of tropical forest succession, even after 200 years (Foster 1990). Saldarriaga et al. (1988) studied 23 stands varying in age from 10 to more than 80 years in the Upper Río Negro of Colombia and Venezuela and found that stands were more homogeneous during the first 10-20 years when pioneer species dominated. At 30-40 years, pioneers were replaced by fast-growing persistent species that dominated over the next 50 years and often persisted in mature forest. At 60-80 years, there were major species changes as large trees died, creating a variety of gap sizes and altering the structure and light availability in the forest.

7.4.1 Functional Groups Affect Successional Patterns

On similar sites, the rates and patterns of secondary succession may be affected by the functional groups represented among the establishing seedlings and saplings. For example, early establishment of a tree cover strongly affects subsequent successional patterns and probably other ecosystem processes as well (Jordan 1985; Ewel et al. 1991). Similarly, early dominance by large-leaved understory herbs, palms, or pioneer herbs and grasses affects microclimates, light environments, and soil conditions, thus affecting establishing seedlings (Denslow 1978; Ewel 1983). However, species diversity within such functional groups may not strongly affect ecosystem processes. Ewel et al. (1991) and Lugo (1992) found that diverse natural successions and tree plantations (monocultures) were similar in their ability to maintain soil fertility, although both were considerably more effective than bare soil.

Rates of succession – changes in biomass, leaf area, species richness, and forest structural heterogeneity – may also be influenced by the number of functional groups in the regeneration pool. For example, rates of succession are rapid following clearcuts and hurricane devastation. Although stand structure and microclimates are strongly modified in both cases, species and functional group diversity are maintained because seed and seedling stocks remain largely intact. Within 1-2 years, strip clearcuts at Palcazu, Peru, were rapidly colonized by most of the tree species present in the original forest (Hartshorn and Pariona 1993). Studies of stand regeneration following Hurricanes Hugo, Gilbert, and Joan in the Caribbean document low mortality rates and rapid reestablishment of leaf area (Bellingham 1991; Walker 1991; Yih et al. 1991). Although it is too early to determine effects on other aspects of stand structure and processes, it appears that the long term impact of hurricanes on the composition of these forests will be less drastic than originally hypothesized.

Structural heterogeneity may also affect successional processes. Saldarriaga et al. (1988) suggest that the more heterogeneous crown structure of old growth tropical forests results in a more heterogeneous light environment at the forest floor (light fleck and gap size-frequency distributions) providing more establishment sites for seedlings than the relatively uniform canopy of a young secondary forest. As a result, rates of seedling establishment are slow while canopy structure is intact during early succession and increase when the cohort of pioneer trees begins to break up. Similar patterns have been described from temperate forests (e.g., Christensen and Peet 1981).

The deviation of some tropical successions from this pattern may be attributed in part to the early dominance of a few functional groups and the absence of others following repeated or intense disturbance. In their survey of secondary forests, Brown and Lugo (1990) observed differences in rates of succession only partially explained by site climatic and edaphic characteristics. Variation in duration and intensity of disturbance affected rates of biomass and soil organic matter accumulation.

Table 7.2. Changes in vegetation structure and diversity during early old field succession in Colombian premontane rain forest (Denslow 1978). The three fields were cleared following different intensities of disturbance: low (secondary forest cleared once 20 years previously), medium (scrub cleared every 3-5 years for the past 20 years), and high (pasture cleared and grazed annually). All plants were counted and identified in 100 1m^2 quadrats at intervals during the first fallow year

Age of fallow (months)	Pasture				Scrub				Secondary forest			
	N	H'$_S$	H'$_G$	S	N	H'$_S$	H'$_G$	S	N	H'$_S$	H'$_G$	S
1					50	3.05	1.30	1.000	40	2.80	1.5	1.000
2	60	3.61	1.35	1.000					58	3.28	1.62	0.733
3					68	3.44	1.34	0.666	72	3.50	1.60	0.575
4	54	3.41	1.38	0.793								
5					73	3.44	1.17	0.579	73	3.52	1.62	0.504
6	47	3.12	1.39	0.700								
7					72	3.57	1.27	0.593				
8									86	3.68	1.64	0.409

[a] N = number of species; H'$_S$ = Shannon-Wiener diversity index based on distribution of individuals among species; H'$_G$ = Shannon-Wiener diversity index based on distribution of individuals among six growth forms (trees, vines and lianas, small herbs, large herbs, small shrubs, large shrubs). S = stand similarity index (2W/A+B) of subsequent stand samples compared to the first sample date and illustrating relative changes in vegetation composition over time (see Greig-Smith 1983 for details on calculations of indices)

Extensive portions of tropical landscapes have been converted from forest to persistent, sometimes exotic grass, fern, or shrub cover following intensive exploitation (Kellman 1980; Woods 1989). These arrested successions are characterized by low biomass and productivity and sometimes by low species richness as well. They are perpetuated by fire and the inability of trees to establish in competition with the existing herbs and shrubs (Uhl et al. 1982; Jordan 1985; Uhl 1987). Where remnant trees persist or, near forest boundaries, grass and shrub growth is limited, tree seedlings are able to establish, and other successional changes are initiated (e.g., Guevara et al. 1986).

Patterns of early succession in fallow fields of traditional agriculturalists illustrate these patterns. In Colombia (Denslow 1978, 1985) and Venezuela (Uhl 1987), composition and structure of fallow fields cleared from old secondary or primary forest changed rapidly during the first year following abandonment. Species diversity increased throughout the first fallow year in Colombia (Table 7.2), with the result that at 8 months the stand was tall (> 2 m) and rich in species as well as life forms. In contrast, fields that had been cleared (cut and burned) annually and occasionally grazed showed little change during the first fallow year. Although initial diversity in these fields was also high, establishing seedlings were displaced by resprouting shrubs and grasses. Vegetation after 6 months was low (ca. 0.5 m tall) dominated by high-light-demanding shrubs and grasses. The inability of seedlings to establish in competition with vegetatively spreading herbs perpetuated high-light environments and the dominance of grasses and herbs for many years.

7.4.2 Causes of Depauperate Regeneration Pools

Several factors may produce regeneration pools deficient in functional groups and species: repeated disturbance, removal of the soil organic layer, biogeographic circumstances, and stringent site conditions.

Repeated disturbance simplifies the patterns and processes of succession. Repeated clearing depletes soil seed stocks and decreases the species diversity of the stand. Canopy removal stimulates the germination of seeds in the surface soils. Soil seed densities decline steeply and do not reach predisturbance levels until rates of seed rain increase with the regrowth of vegetation on the site (Young et al. 1987). Even then, however, the composition of the soil seed stocks is heavily dominated by pioneer species, whereas the seed stocks under both old secondary and primary forest are rich in species and lifeforms (Uhl et al. 1982; Young et al. 1987).

Repeated clearing, burning, or grazing selectively eliminates those species that do not resprout or that have not reached reproductive maturity in the interim (e.g., some palms and many understory and shade tolerant canopy trees; Uhl 1987). Moreover, new arrivals in the seed rain or from the soil stocks are displaced by the vegetative growth of species and functional groups capable of resprouting or clonal spread (e.g., shrubs, grasses, vines). As a result, successions on repeatedly cleared land may be missing some functional groups, although they may be rich in species of high-light-demanding herbs and shrubs.

Removal of the soil organic layer (often thin in tropical soils) through landslides or the impact of logging machinery (e.g., skid tracks and logging roads), also removes seed stocks and mycorrhizal innoculum. Species with copious, well-dispersed seeds (pioneer trees and high-light-demanding shrubs, vines, and herbs) and facultatively mycorrhizal species can dominate these sites more rapidly than species with less well dispersed seeds or for which seedling growth depends on mycorrhizal development (Janos 1980).

Some functional groups may be absent due to circumstances of biogeography or history. For example, islands may be dominated by well-dispersed, high-light-demanding species with few large-seeded, shade-tolerant canopy emergents or understory species (Vitousek et al, 1995). The dominance of *Metrosideros* in Hawaiian montane rain forest is an example.

Site conditions may prevent establishment of some groups. High water tables, low nutrient-supply rates, saline environments, and long dry seasons affect some groups more than others. Diversity of both species and functional groups are likely to be lower in such sites.

7.5 Functional Groups in Tropical Dry Forests

Stature, species richness, and structural complexity in general decrease along a gradient of declining annual rainfall or increasing seasonality, although the pattern is affected by edaphic and biogeographic circumstances as well (Beard 1955; Sarmiento 1972; Gentry 1995). Although tropical dry forsts (TDF) share few species with tropical moist forests (TMF), most genera and families appear to be of TMF origin. Lower species diversity in TDF is largely the result of loss of these TMF genera (Gentry 1995) and some functional groups are better represented in tropical dry forests than others. Palms, epiphytes, and large-leaved understory herbs and shrubs such as Araceae are scarce in all but the wettest TDF. Legumes remain an important component of the canopy tree flora in Neotropical TDF. Liana diversity is low in TDF although stem density remains high. In contrast, the richness of vines increases as does the richness of understory herbs and shrubs (Gentry 1995). As forest stature declines with decreasing rainfall, the abundance of emergent trees decreases (Beard 1955). Whereas most TMF trees and shrubs are bird or mammal dispersed, in TDF wind dispersal is prevalent (Gentry 1995). Consequently, the abundance of frugivorous birds declines as well (Stiles 1983).

Natural disturbance regimes of TDF are similar to those of TMF. Treefalls are the primary source of small-scale canopy openings and some tropical forests are subject to periodic hurricanes and typhoons as well (Quigley 1994). Although grassland fires (pasture or savanna) burn into forest edges (Murphy and Lugo 1986) and often affect logged- or hurricane-damaged forest (Whigham et al. 1991), there is little evidence that closed canopy forests burn naturally in the absence of human intervention (Murphy and Lugo 1986). Today, however, successions following forest clearing or conversion to pasture are effectively arrested by frequent grass fires that destroy tree and shrub regeneration.

Leaf area indices in TDF are often less than half of those in TMF (Murphy and Lugo 1986) due to fewer canopy strata, prevalence of drought deciduous species, and small leaf sizes (Medina 1995). Abbreviated growing seasons reduce net primary productivity as well, with the result that canopy openings are slower to close and light levels in the understory are higher than in TMF. Understory shrub and herb densities are correspondingly high (Gentry 1995), and seedling growth and establishment may be less limited by light than by moisture availability (Lott et al. 1987; Quigley

1994). In forests disturbed frequently by hurricanes, the canopy is broken and open, regeneration is abundant, and reproduction is common (Quigley 1994). Although regrowth and canopy closure are slow, the composition of early and late successional communities in TDF environments is often similar both because late successional species are tolerant of hot dry conditions and because resprouting (coppicing) is common. However, because plant growth is slow in tropical dry forest environments, forest structure (including basal area, stem densities, height) retains the effects of clearing for a longer time than does rain forest. In this respect, TDF may be more vulnerable to human impacts than is tropical rain forest.

Table 7.3. Species diversity within functional groups in two Neotropical rain forest floras. Barro Colorado Island, Panama, a seasonally deciduous tropical moist forest, receives about 2750 mm annual precipitation; rainfall is less than 100 mm during 4 months of the year (Croat 1978). La Selva, a premontane wet tropical forest, receives about 3960 mm of annual rainfall with no month having less than 100 mm (Wilbur 1994)

Group	BCI, Panama		La Selva, CR	
	S	%[a]	S	%[a]
A. Herbs and shrubs				
1. Pioneer	252	21.1	349	20.8
2. Understory				
a. Broad-leaved	45	3.8	88	5.2
b. Small-leaved	137	11.4	270	16.1
B. Treelets				
1. Pioneer	43	3.6	80	4.8
2. Understory	176	14.7	172	10.2
C. Canopy trees[b]	91	7.6	168	10.2
1. Legumes	18	1.5	43	2.6
2. Palms	3	0.2	7	0.4
D. Emergents[c]	28	2.5	13	0.8
E. Lianas and vines				
1. Lianas	150	12.5	100	6.0
2. Vines	120	10.0	72	4.3
F. Epiphytes	154	12.9	367	21.8
Species classified[d]	1196		1679	
Species in flora[e]	1369		1813	

[a]Percent classified species.
[b]Canopy trees include legumes and palms as well as other species which regularly constitute the main canopy of the forest. Pioneer trees, also part of the main canopy following disturbance, are not included in this category.
[c]Emergent legumes were classifed as emergents for this table.
[d]Lycopodiaceae, Selaginaceae, and cultivated plants are not included in this classification. Others were not included because the growth form was unknown.
[e]Croat (1978); Wilbur (1994).

It is difficult to generalize about either structure or relative abundance of different functional groups in tropical dry forests because both change dramatically along rainfall, seasonality, and edaphic gradients. At the moist end of the gradient, light is still an important limitation to plant growth and the functional groups identified for tropical rain forest affect regeneration processes in more seasonal tropical moist forests as well.

The distribution of species among functional groups is similar at BCI and La Selva, although with some notable differences (Table 7.3). At La Selva, Costa Rica, an aseasonal tropical wet forest, relative and absolute diversity of understory herbs and shrubs, canopy trees, and epiphytes is higher than at Barro Colorado Island, a seasonally deciduous tropical forest, where understory treelets, emergent trees, and lianas and vines are more diverse. Although species richness does not necessarily reflect abundance, these differences in the distribution of diversity are consistent with observed differences in structure of the two forests.

These differences in development of functional groups may lead to differences in disturbance ecology as well. Seedling growth should be less limited by light at BCI than at La Selva because the understory may be more open (fewer understory herbs and shrubs) and canopy gaps larger (more emergents and lianas) than at La Selva. We expect that gaps will close more slowly because of lower growth rates in seasonally dry climates, also contributing to high light availabilities in the understory. However, we also expect that seedling establishment and growth will be more limited by moisture availability. Data from field observations suggest that such is the case: seedling establishment and plant growth are strongly affected by light availability (Augspurger 1984; Brokaw and Scheiner 1989; Smith 1987), seedling establishment is more strongly seasonal at BCI than in wetter forests (Garwood 1983), and spatial heterogeneity in seedling abundance reflects patterns of moisture availablity as well as light availability (Becker et al. 1988; Hubbell, unpubl.).

Along a gradient of declining rainfall and increasing seasonality, then, some functional groups are lost and moisture becomes a more important limiting factor than light. Because moisture supply is dependent on edaphic as well as climatic conditions, spatial heterogeneity in soil depth, proximity to water courses, bedrock structure, etc. may have a stronger effect on dry forest structure and composition than in TMF (Medina 1995). Likewise, definitions of functional groups for these forests should incorporate their effects on and response to moisture availability. For example, distributions of fine and coarse roots and foliage characteristics (deciduousness, phenology, photosynthetic pathway, and canopy architecture) will be important defining characteristics (see Bullock et al. 1995, for more detail).

7.6 Redundancy within Functional Groups

Although species within functional groups are similar in their effects on successional processes, they differ in their habitat requirements, reproductive characteristics, foliage quality, and susceptibility to insect and pathogen damage. Species diversity within functional groups may produce robustness in ecosystem processes (Ewel et al. 1991; Lawton and Brown 1993; Vitousek and Hooper 1993) in the sense that disturbance regimes, nutrient turnover rates, and patterns of regrowth may be less temporally and spatially variable in species-rich than in species-poor ecosystems.

At La Selva, two species of emergent trees in the family Leguminosae have approximately reciprocal distributions. *Dipteryx panamensis* is restricted to alluvial soils and *Pithecellobium elegans* occurs almost exclusively on the poorer, residual soils of the uplands. Another emergent, *Carapa guianensis* (Meliaceae), is abundant on swampy soils (Clark 1994). Their great sizes result in large canopy openings when they fall, producing similar impacts on the disturbance regimes of forest on all three soil types.

Similarly, the canopies of early successional forests are often dominated by one or a few species of pioneer trees. Within the same region, however, stands of the same age and similar history may be dominated by different species, perhaps because of differences in disturbance history, seed availability, soil structure, or season in which the site became available. Nevertheless, successional processes in tree-dominated stands most likely follow similar patterns of cohort establishment, stand thinning, and gap formation (Saldarriaga et al. 1988; Brown and Lugo 1990), and have similar effects on soil processes (Lugo 1992; Ewel et al. 1991).

Perhaps because of this robustness, tropical rain forests are not dominated by keystone species (sensu Gilbert 1980) that effectively control ecosystem processes. However, the dependence of this species diversity on a relatively narrow climatic range of high temperatures and high, aseasonal annual rainfall may make tropical rain forest dynamics particularly susceptible to climate change.

7.7 Conclusions

Functional groups may be usefully defined according to their effect on a particular pattern or process. The categories proposed here with respect to patterns of disturbance and regeneration in tropical rain forest are neither exclusive nor exhaustive. They are constructed to recognize groups of species that have similar impacts on forest dynamics. Other groups might be defined according to the effects of species on nutrient cycling (e.g., due to foliage quality and nutrient content) or net primary productivity (e.g., based on photosynthetic light responses or carbon allocation patterns). If we are to understand the consequences of diversity for important eco-system processes, it will be important explicitly to recognize the ways in which different groups of species affect those processes.

In tropical moist forests where light is an important factor limiting plant growth, canopy opening events and revegetation patterns are affected by the relative abundances of different functional groups that in some way affect the spatial and temporal distribution of light. In tropical dry forests, disturbance regimes and stand dynamics may be strongly influenced by spatial and temporal distributions of moisture supply as well. In these forests, the definition of functional groups affecting stand dynamics should take into account differences in responses to moisture availability and effects on moisture supply.

Species in different functional groups not only differently affect disturbance and regrowth, they are also differentially vulnerable to human impacts. In rain forests, understory shrubs and epiphytes are slower to recover following forest clearing than are either canopy or emergent trees. Identification of functional groups may facilitate prediction of the consequences of human exploitation to particular groups of organisms as well as the consequences of species loss for disturbance and regeneration processes.

Acknowledgements. I am grateful to the National Science Foundation for support for studies on tropical rain forest dynamics and succession to the Organization for Tropical Studies (BSR83-06923, BSR86-05106, and DEB92-08031) and to OTS for providing the environment in which to develop long term research on forest processes. The manuscript was improved by discussion with the Tropical Forestry Group at the University of Florida, and with the Oaxtepec Workshop on the Impact of Biodiversity on Ecosystem Processes in Tropical Forests and from critical comments from N. Brokaw, H. Cushman, J. Ewel, C. S. Holling, J. Putz, G. Orians, and V. Viana.

References

Alvarez-Buylla ER, Garcia-Barrios R (1991) Seed and forest dynamics: a theoretical framework and an example from the Neotropics. Am Nat 137:133-154

Augspurger CK (1984) Pathogen mortality of tropical tree seedlings: experimental studies of the effects of dispersal distance, seedling density, and light conditions. Oecologia 61:211-217

Augspurger CK, Hutchings MW, Watson AR, Davy AJ (1988) Impact of pathogens on natural plant populations. In: Plant population ecology. 28th Symp Br Ecol Soc, Blackwell, pp 413-433

Bazzaz FA, Pickett STA (1980) Physiological ecology of tropical succession: a comparative review. Ann Rev Ecol Syst 11:287-310

Beard JS (1955) The classification of tropical American vegetation types. Ecology 36:89-100

Becker P, Rabenold PE, Idol JR, Smith AP (1988) Water potential gradients for gaps and slopes in a Panamanian tropical moist forest dry season. J Trop Ecol 4:173-184

Bellingham PJ (1991) Landforms influence patterns of hurricane damage: evidence from Jamaican montane forests. Biotropica 23:427-433

Blair JM, Parmalee RW, Beare MH (1990) Decay rates, nitrogen fluxes, and decomposer communities of single- and mixed-species foliar litter. Ecology 71:1976-1985

Bloomfield J, Vogt KA, Vogt DJ (1993) Decay rate and substrate quality of fine roots and foliage of two tropical tree species in the Luquillo Experimental Forest, Puerto Rico. Plant and Soil 150:233-245

Brandani A, Hartshorn GS, Orians GH (1988) Internal heterogeneity of gaps and species richness in Costa Rican tropical wet forest. J Trop Ecol 4:99-119

Brokaw NVL (1982) Treefalls: frequency, timing and consequences. In: Leigh Jr EG, Rand AS, Windsor DM (eds) The ecology of a tropical forest: seasonal rhythms and long-term changes. Smithsonian Inst Press, Washington DC, pp 101-108

Brokaw NVL (1985) Treefalls, regrowth and community structure in tropical forests. In: Pickett STA, White PS (eds) The ecology of natural disturbance and patch dynamics. Academic Press, Orlando, pp 53-69

Brokaw NVL, Scheiner SM (1989) Species composition in gaps and structure of a tropical forest. Ecology 70:538-514

Brown S, Lugo AE (1990) Tropical secondary forests. J Trop Ecol 6:1-32

Bullock SH, Mooney HA, Medina E (1995) Seasonally dry tropical forests. Cambridge Univ Press, New York

Christensen NL, Peet RK (1981) Secondary forest succession on the North Carolina piedmont. In: West DC, Shugart HH, Botkin DB (eds) Forest succession: concepts and applications. Springer, Berlin Heidelberg New York, pp 230-245

Clark DA (1994) Plant demography. In: McDade LA, Bawa KS, Hespenheide HA, Hartshorn GS (eds) La Selva. Ecology and natural history of a Neotropical rain forest. Univ Chicago Press. Chicago, pp 90-105

Clark DA, Clark DB (1992) Life history diversity of canopy and emergent trees in a neotropical rainforest. Ecol Monogr 62:315-344

Clark DB (1991) The impact of physical damage on canopy tree regeneration in tropical rainforest. J Ecol 79:447-457

Condit R, Hubbell SP, Foster RB (1992) Recruitment near conspecific adults and the maintenance of tree and shrub diversity in a neotropical forest. Am Nat 140:261-286

Connell JH (1978) Diversity in tropical rain forests and coral reefs. Science 199:1302-1310

Connell JH, Lowman MD (1989) Low-diversity tropical rain forests: some possible mechanisms for their existence. Am Nat 134:88-119

Connell JH, Slatyer RO (1977) Mechanisms of succession in natural communities and their role in community stability and organization. Am Nat 111:1119-1144

Croat TB (1978) Flora of Barro Colorado Island. Stanford Univ Press, Stanford, CA

Crow TR (1971) A rain forest chronicle: 30 year record of change in structure and composition at El Verde, Puerto Rico. Biotropica 12:42-55

Denslow JS (1978) Secondary succession in a Colombian rainforest: Strategies of species response across a disturbance gradient. PhD Thesis, Univ Wisconsin, Madison

Denslow JS (1980a) Gap partitioning among tropical rainforest trees. Biotropica 12(Suppl):47-55

Denslow JS (1980b) Patterns of plant species diversity during succession under different disturbance regimes. Oecologia 46:18-21

Denslow JS (1985) Disturbance-mediated coexistence of species In: Pickett STA, White PS (eds) The ecology of natural disturbance and patch dynamics. Academic Press, Orlando, pp 307-323

Denslow JS (1987) Tropical treefall gaps and tree species diversity. Ann Rev Ecol Syst 18:431-451

Denslow JS, Hartshorn GS (1994) Treefall gap environments. In: McDade LA, Bawa K, Hespenheide H, Hartshorn GS (eds) La Selva: ecology and natural history of a Neotropical rain forest. Univ Chicago Press, Chicago, pp 120-127

Denslow JS, Schultz J, Vitousek PM, Strain B (1990) Growth response of tropical shrubs to treefall gap environments. Ecology 71:165-179

Denslow JS, Newell E, Ellison AM (1991) The effect of understory palms and cyclanths on the growth and survival of Inga seedlings. Biotropica 23:225-234

Drury WH, Nisbet ICT (1973) Succession. J. Arnold Arbor 16:331-368

Ellison AM, Denslow JS, Loiselle BA, Brenes D (1993) Seed and seedling ecology of Neotropical Melastomataceae. Ecology 74:1733-1749

Ewel JJ (1983) Succession. In: Golley FB (ed) Tropical rainforest ecosystems. Elsevier, Amsterdam, pp 217-223

Ewel JJ, Mazzarino MJ, Berish CW (1991) Tropical soil fertility changes under monocultures and successional communities of different structure. Ecol Appl 1:289-302

Foster RB (1990) Long-term change in the successional forest community of the Río Manu floodplain. In: Gentry AH (ed) Four Neotropical forests. Yale Univ Press, New Haven, pp 565-572

Garwood N (1983) Seed germination in a seasonal tropical forest in Panama: a community study. Ecol Monogr 53:159-181

Garwood N, Janos DP, Brokaw N (1979) Earthquake-caused landslides: a major disturbance to tropical forests. Science 205:997-999

Gentry AH (1982) Patterns of neotropical plant species diversity. Evol Biol 15:1-84

Gentry AH (1995) Diversity and floristic composition of Neotropical dry forests. In: Bullock SH, Mooney HA, Medina E (eds) Seasonally dry tropical forests. Cambridge Univ Press, Cambridge

Gilbert LE (1980) Food web organization and conservation of Neotropical diversity. In: Soulé ME, Wilcox B (eds) Conservation biology. Sinauer, Sunderland, MA, pp 11-33

Goldhammer JG, Seibert B (1989) Natural rain forest fires in eastern Borneo during the Pleistocene and Holocene. Naturwissenschaften 76:518-520

Greig-Smith P (1983) Quantitative plant ecology, 3rd edn. Univ California Press, Berkeley, CA

Grime JP (1977) Evidence for the existence of three primary strategies in plants and its relevance to ecological and evolutionary theory. Am Nat 111:1169-1194

Guariguata MR (1990) Landslide disturbances and forest regeneration in upper Luquillo Mountains of Puerto Rico. J Ecol 78:814-832

Guevara S, Purata SE, van der Maarel E (1986) The role of remnant forest trees in tropical secondary succession. Vegetatio 66: 77-84

Hallé F, Oldeman RAA, Tomlinson B (1978) Tropical trees and forests. Springer, Berlin Heidelberg New York.

Hart TB, Hart JA, Murphy PG (1989) Monodominant and species-rich forests of the humid tropics: causes for their co-occurrence. Am Nat 133:613-633

Hartshorn GS, Pariona W (1993) Ecological sustainable forest management in the Peruvian Amazon. In: Potter C, Cohen J, Janczewski D (eds) Perspectives on biodiversity: case studies of genetic resource conservation and development. AAAS Press, Washington DC, pp 151-166

Holling CS, Schindler DW, Walker BW, Roughgarden J (1995) Biodiversity in the functioning of ecosystems: an ecological primer and synthesis. In: Maler KG, Floke C, Jansson BO, Holling CS (eds) Biodiversity loss: theoretical economic and ecological issues (in press)

Howe HF, Smallwood J (1982) Ecology of seed dispersal. Ann Rev Ecol Syst 13:201-228

Hubbell SP, Foster RB (1986a) Commonness and rarity in a Neotropical forest: implications for tropical tree conservation. In: Soulé M (ed) Conservation biology: science of scarcity and diversity. Sinauer, Sunderland, MA, pp 205-231

Hubbell SP, Foster RB (1986b) Biology, chance and history and the structure of tropical rain forest tree communities. In: Diamond J, Case TJ (eds) Community ecology. Harper and Row, New York pp 314-329

Huston M (1979) A general hypothesis of species diversity. Am Nat 113:81-101

Huston MA, Smith TM (1987) Plant succession: life-history and competition. Am Nat 130:168-198

Janos DP (1980) Mycorrhizae influence succession. Biotropica 12 (Suppl):56-64

Janzen DH (1970) Herbivores and the number of tree species in tropical forests. Am Nat 104:501-528

Jordan CF (1985) Nutrient cycling in tropical forest ecosystems. Wiley, New York

Kellman M (1980) Geographic patterning in tropical weed communities and early secondary successions. Biotropica 12:34-39

Kellman M, Adams CD (1970) Milpa week's of Cayo District, Belize (British Honduras). Can Geogr 14:323-343

Knight DH (1975) A phytosociological analysis of species rich tropical forest on Barro Colorado Island, Panama. Ecol Monogr 45:259-284

Kowal NE (1966) Shifting cultivation, fire, and pine forest in the Cordillera Central, Luzon, Philippines. Ecol Monogr 36:389-419

Lawton JH, Brown VK (1993) Redundancy in ecosystems. In: Schulze E-D, Mooney HA (eds) Biodiversity and ecosystem function. Ecological Studies 99. Springer, Berlin Heidelberg New York, pp 255-270

Lieberman D, Lieberman M, Hartshorn GS, Peralta R (1985) Growth rates and age-size relationships of tropical wet forest trees in Costa Rica. J Trop Ecol 1:97-109

Loiselle BA, Blake JG (1991) Temporal variation in birds and fruits along an elevational gradient in Costa Rica. Ecology 72:180-193

Lott EJ, Bullock SH, Solis-Magallanes JA (1987) Floristic diversity and structure of upland and arroyo forests of coastal Jalisco. Biotropica 19:228-235

Lugo A (1992) Comparison of tropical tree plantations with secondary forests of similar age. Ecol Monogr 62:1-41

Medina E (1995) Diversity of life-forms of higher plants in Neotropical dry forests. In: Bullock SH, Mooney HA, Medina E (eds) Seasonally dry tropical forests. Cambridge Univ Press, Cambridge

Moermond TC, Denslow JS (1985) Neotropical frugivores: patterns of behavior, morphology, and nutrition with consequences for fruit selection. In: Buckley PA, Foster MS, Morton ES, Ridgely RS, Smith NG (eds) Neotropical ornithology. AOU Monogr, vol 36, pp 865-897

Mueller-Dombois D (1983) Canopy dieback and successional processes in Pacific forests. Pac Sci 37:317-325

Murphy PG, Lugo AE (1986) Ecology of tropical dry forest. Ann Rev Ecol Syst 17:89-96

Nye PH, Greenland P (1961) The soil under shifting cultivation. Commonw Agric Bur, Farnham, England

Pickett STA (1976) Succession: an evolutionary interpretation. Am Nat 110:107-119

Putz FE (1984) The natural history of lianas on Barro Colorado Island, Panama. Ecology 65:1713-1724

Putz FE, Mooney HA (1991) The biology of vines. Cambridge U. Press, New York

Quigley MF (1994) Latitudinal gradients in temperate and tropical seasonal forests. PhD Dissertation, Louisiana State Univ, Baton Rouge, LA

Rankin JM (1978) The influence of seed predation and plant competition on tree species abundances in two adjacent tropical rain forests. PhD Thesis, Univ Michigan, Ann Arbor

Rankin-de-Merona JM, Hutchings H, RW, Lovejoy TE (1990) Tree mortality and recruitment over a five year period in undisturbed upland rain forest of the Central Amazon. In: Gentry AH (ed) Four Neotropical rainforests. Yale Univ Press, New Haven, pp 573-584

Raunkiaer C (1934) The life-forms of plants and statistical plant geography. Oxford Univ Press, Oxford

Read DJ (1991) Mycorrhizas in ecosystems. Experientia 47:376-390

Redford KH, Padoch C (eds) (1992) Conservation of Neotropical rainforests: working from traditional resource use. Columbia Univ Press, New York

Root R (1967) The niche exploitation pattern of the blue-grey gnatcatcher. Ecol Monogr 37:317-350

Saldarriaga JG, Luxmoore RJ (1991) Solar energy conversion efficiencies during succession of a tropical rain forest in Amazonia. J Trop Ecol 7:233-242

Saldarriaga JG, West DC (1986) Holocene fires in the north Amazon Basin. Quat Res 26:358-366

Saldarriaga JG, West DC, Tharp ML, Uhl C (1988) Long term chronosequence of forest succession in the upper Rio Negro of Colombia and Venezuela. J Ecol 76:938-958

Salo J, Kalliola R, Häkkinen I, Mäkinen Y, Niemelä P, Puhakka M, Coley PD (1986) River dynamics and the diversity of Amazon lowland forest. Nature (Lond) 322:254-258

Sanford Jr. RL, Baker HE, Hartshorn GS (1986) Canopy openings in a primary Neotropical lowland forest. J Trop Ecol 2:277-282

Sanford Jr RL, Saldarriaga J, Clark KE, Uhl C, Herrera R (1985) Amazon rainforest fires. Science 227:53-55

Sarmiento G (1972) Ecological and floristic convergences between seasonal plant formations of tropical and subtropical South America. J Ecol 60:367-410

Smith AP (1987) Respuestas de hierbas del sotobosque tropical a claros ocasionados por la caìda de àrboles. Rev Biol Trop 35(Suppl 1):111-118

Solbrig OT (1993) Plant traits and adaptive strategies: their role in ecosystem function. In: Schulze E-D, Mooney HA (eds) Biodiversity and ecosystem function. Ecological Studies 99. Springer, Berlin Heidelberg New York, pp 97-116

Sollins P, Radulovich R (1988) Effects of soil physical structure on solute transport in a weathered tropical soil. Soil Sci Soc Am J 52:1168-1173

Stemmermann L (1983) Ecological studies of Hawaiian *Metrosideros* in a successional context. Pac Sci 37:361-373

Stiles FG (1983) Chapter 10. Birds. In: Janzen DH (ed) Costa Rican natural history. Univ Chicago Press, Chicago, Illinois, pp 502-618

Stiles FG (1988) Altitudinal movements of birds on the Caribbean slope of Costa Rica: implications for conservation. In: Almeda F, Pringle CM (eds) Tropical rain forest: diversity and conservation. Calif Acad Sci, San Francisco, pp 243-258

Strong D (1977) Epiphyte loads, treefalls, and perennial forest disruption: a mechanism for maintaining higher tree species richness in the tropics without animals. J Biogeogr 4:215-218

Uhl C (1987) Factors controlling succession following slash-and-burn agriculture in Amazonia. J Ecol 75:377-407

Uhl C (1988) Restoration of degraded lands in the Amazon basin. In: Wilson EO, Peter FM (eds) Biodiversity. Natl Acad Press, Washington DC, pp 326-333

Uhl C, Kauffman JB (1990) Deforestation, fire susceptibility, and potential tree responses to fire in the eastern Amazon. Ecology 71:437-449

Uhl C, Clark H, Clark K (1982) Successional patterns associated with slash-and-burn agriculture in the upper Rio Negro region of the Amazon Basin. Biotropica 14:249-254

Uhl C, Clark K, Dezzeo N, Maquirino P (1988) Vegetation dynamics in Amazonian treefall gaps. Ecology 69:751-763

Vitousek PM (1984) Litterfall, nutrient cycling, and nutrient limitation in tropical forests. Ecology 65:285-298

Vitousek PM, Denslow JS (1986) Nitrogen and phosphorus availability in treefall gaps of a lowland tropical rain forest. J Ecol 74:1167-1178.

Vitousek PM, Hooper DM (1993) Biological diversity and terrestrial ecosystem biogeochemistry. In: Schulze E-D, Mooney HA (eds) Biodiversity and ecosystem function. Ecological Studies 99. Springer, Berlin Heidelberg New York, pp 3-14

Vitousek PM, Walker LR (1989) Biological invasion of *Myrica faya* in Hawaii: plant demography, nitrogen fixation, and ecosystem effects. Ecol Monogr 59:247-265

Vitousek PM, Walker LR, Whiteaker LD, Mueller-Dombois D, Matson PA (1987) Biological invasion by *Myrica faya* alters ecosystem development in Hawaii. Science 238:802-804

Vitousek PM, Adsersen H, Loope LL (1995) (eds) Islands: biologica, diversity and ecosystem function. Springer, Berlin Heidelberg New York

Walker LR (1991) Tree damage and recovery from Hurricane Hugo in Luquillo Experimental forest, Puerto Rico. Biotropica 23:379-385

Weaver PL, Birdsey RA, Lugo AE (1987) Soil organic matter in secondary forests of Puerto Rico. Biotropica 19:17-23

Wheelwright NT (1985) Fruit size, gape width, and diets of fruit- eating birds. Ecology 66:808-818

Whigham DF, Olmsted I, Cabrera Cano E, Harmon ME (1991) The impact of Hurricane Gilbert on trees, litterfall, and woody debris in a dry tropical forest in the northeastern Yucatan Peninsula. Biotropica 23:434-441

Whitmore TC (1984) Tropical rain forests of the Far East. Clarendon Press, Oxford

Whitmore TC (1989) Changes over 21 years in the Kolombangara rainforests. J Ecol 77:469-483

Whitmore TC, Sayer JA (1992) Tropical deforestation and species extinction. Chapman and Hall, London

Wilbur RL, Collaborators (1994) Vascular Plants: An interim checklist. In: McDade LA, Bawa K, Hespenheide H, Hartshorn GS (eds) La Selva: ecology and natural history of a Neotropical rain forest. Univ Chicago Press, Chicago, pp 350-378

Wilson EO (ed) (1988) Biodiversity. Natl Acad Press, Washington DC

Woods P (1989) Effects of logging, drought, and fire on structure and composition of tropical forests in Sabah, Malaysia. Biotropica 21:290-298

Yih K, Boucher DH, Vandermeer JH, Zamora N (1991) Recovery of rain forest of southeastern Nicaragua after destruction by Hurricane Joan. Biotropica 23:106-113

Young KR, Ewel JJ, Brown BJ (1987) Seed dynamics during forest succession in Costa Rica. Vegetatio 71:157-173

Young TP, Hubbell SA (1991) Crown asymmetry, treefalls, and repeat disturbance of broadleaved forest gaps. Ecology 72:1464-1471

8 Species Richness and Resistance to Invasions

Marcel Rejmánek[1]

8.1 Diversity vs. Stability

Traditionally, tropical forests, and especially tropical rain forests, have been contrasted with extratropical communities in terms of their species diversity and stability (Elton 1958). Unfortunately, ecologists have used the word 'stability' to mean several different things (Orians 1975; Harrison 1979; Pimm 1984): *Resilience* can be defined as a rate of return of population densities, community composition, or collective properties like biomass production, to conditions preceding a perturbation. *Persistence* usually means how long presence of individual populations or community composition last. *Resistance* means the degree to which a variable of interest remains unaltered following perturbation. *Constancy* usually means a lack of change (low variability) of variables of interest over time. Finally, systems are defined as *stable in a narrow sense* if, and only if, variables of interest return to their initial (equilibrium) values, following perturbation. Elton (1958) himself switched between different meanings of stability when he talked about absence of insect outbreaks in tropical forests (high population *constancy*) and about higher frequency of extinctions and invasions in simple communities (low *persistence*). Elton suggested that species rich communities like tropical rain forests possess "complex systems of checks and buffers" responsible for their stability. Causal positive connections between biotic diversity and low variability or high persistence of tropical comunities have been questioned many times since Elton's influential book was published (Futuyma 1973; Farnworth and Golley 1974; Leigh 1975; Wolda 1978, 1983; Maury-Lechon et al. 1984). Elton, however, should be prized for drawing the long-lasting attention of ecologists to relationships between diversity and stability in ecological systems.

After the probability of stability (in narrow sense) of ecological systems was first defined in mathematical terms (MacArthur 1970; May 1973), theoretical results seemed to contradict Elton's generalizations. Attempts were

---

[1] Department of Botany, University of California, Davis, Davis, California 95616, USA

Ecological Studies, Vol. 122
Orians, Dirzo and Cushman (eds) Biodiversity and Ecosystem Processes in Tropical Forests
© Springer-Verlag Berlin Heidelberg 1996

then made to estimate values of relevant parameters in simple, real biotic communities (Seifert and Seifert 1976; McNaughton 1978; Rejmánek and Stary 1979). Progress, however, has been frustratingly slow (Pimm 1984; Hallett 1991; Rejmánek 1992; Roxburgh 1994) and our understanding of relationships between species diversity and different types of stability remains rudimentary at best. The only conclusion emerging from empirical studies so far is that some functional properties of ecosystems, such as rate of decomposition (Springett 1976; Hobbs 1992), productivity (Naeem et al. 1994), nutrient cycling (Vitousek and Hooper 1993), tolerance of herbivory (Brown and Ewel 1988; McNaughton 1993), or drought resistance (Tilman and Downing 1994), seem to be positively related to species diversity, at least over some range of species numbers. It seems that higher species diversity brings higher functional diversity and, thanks to that, also higher resistance and/or resilience of some collective properties of ecosystems. Whether the relationship is truly monotonic remains to be seen.

Resistance to invasions of alien species has been recognized as one measurable kind of ecological stability or, more precisely, of composition persistence (Elton 1958; Rejmánek 1989; Trepl 1990; Pimm 1993). At least some theoretical studies show that the probability of invasion success should decrease with species number and with strength of interspecific interactions in multispecies systems (Robinson and Valentine 1979; Post and Pimm 1983; Drake 1988; Case 1990, 1991). One of the reasons why the large or strongly connected model competition communities repel invaders is the emergence of multiple domains of attraction in such systems. The existence of multiple domains of attraction in multispecies systems always gives the disadvantage to species which, although perhaps equally competitive with the others, are latecomers and at low frequency (Case 1991). Case's reference to 'activation barriers' produced in complex model communities echoes the 'biotic bariers' in complex communities anticipated by Allee et al. (1949). According to food-web assembly simulations by Post and Pimm (1983) and Drake (1988), model communities are easier to invade early in the assembly process when they have few species and when they have simple patterns of interspecific interactions. The question, however, still remains whether there is really any relationship between species diversity and resistance to invasions in real biotic communities as proposed by Allee et al. (1949) and Elton (1958). Are species-rich, tropical communities truly more invasion resistant than species-poor temperate communities? Unfortunately, even recently published papers on tropical invasions strongly disagree as to whether communities in the tropics are more or less prone to invasions than communities in extratropical areas (Usher 1991; Whitmore 1991).

This chapter addresses the following questions:

1. Do the tropics (and tropical forests in particular), accumulate smaller numbers (or percentages) of non-native species relative to extratropical ecosystems?

2. If so, is such a difference really an expression of greater resistance to invasions?

3. If so, is the high species richness of native biota at least partly responsible for this resistance?

8.2 Global Patterns

The attempt to quantify global latitudinal patterns of species richness in native and established non-native biota has, surprisingly, apparently never been undertaken even for the best known groups such as vascular plants.

Species richness of alien biotas in a particular area can be expressed in three different ways: (1) The total number of alien species, (2) the percentage of alien species, and (3) the number of alien species/log(area). The total number of alien species is not particularly revealing by itself, but becomes informative when scaled in relation to the area under consideration. The percentage of alien species in the flora or fauna is a useful index, but may be biased by low or high richness of native biota. The number of alien species/log(area) can be used as a standardized expression of alien species richness because there is generally an approximately linear relationship between the number of species in an area and log(area). When log to the base 10 is used, this index corresponds to the extrapolated mean number of alien species per 10 km^2. For the purposes of this analysis, the latter two ways of expressing the diversity of alien floras will be used.

Percentages of naturalized alien vascular plant species in 38 continental areas of the Americas and 14 continental areas of Europe and Africa were determined from recently published regional floras and checklists. Figure 8.1A presents these percentages as a function of latitude. The relationship is bimodal with maxima at 45°N and 40°S, a pattern that could result from the fact that numbers of native species/log(area) peaks (with the exception of South Africa) in the equatorial tropics (Fig. 8.1C). However, the number of alien species/log(area) as a function of latitude is also clearly bimodal (Fig. 8.1B) with maxima at about 40°N and 35°S, and a depression in the equatorial tropics. Available data from Australia concur with these trends (Clarkson and Kenneally 1988; Kloot 1991; Cowie and Werner 1993). Collectively, these data offer a partial answer to our first question: yes, the continental tropics have accumulated lower numbers (and percentages) of non-native species of vascular plants than temperate areas have.

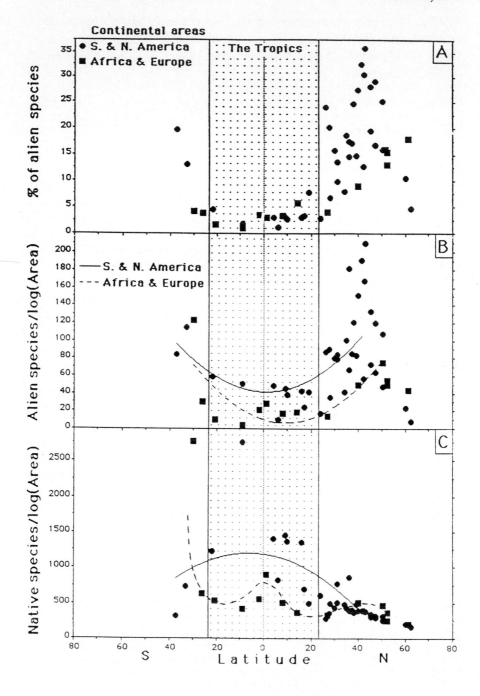

Vascular plant floras for 63 oceanic islands were analyzed in the same way; the results are presented in Fig. 8.2. Global trends for alien species on islands are dramatically different from those in continental areas: although the highest values of percentages of alien species occur within the tropics, there is no statistically distinguishable pattern (with the exception of trivial declines towards polar areas). Based on these data, it can be concluded that numbers of non-native species are about the same on tropical islands as they are on temperate islands.

There are three, mutually nonexclusive, suites of explanations for these patterns:

1. Differences in the amount of disturbance by humans (continental tropical vs. continental extratropical areas, continents vs. islands).

2. Differences in the quantity and quality of imported propagules of alien plants (e.g., mainly 'wrong' temperate species have been imported to the tropics).

3. Differences in invasion resistance between continental tropical areas and continental extratropical areas due to natural biotic and/or abiotic factors.

In general, plants are more likely to become invasive in human-disturbed habitats than in relatively pristine habitats (Rejmánek 1989; Hobbs and Huenneke 1992). Surprisingly, however, tropical areas like Burma, Senegal, Java, or Cuba, which are as highly disturbed as temperate areas (Hannah et al. 1994), also have rather low numbers of alien plant species/log(area): 34.3 (Hundley and Chit 1961), 22.9 (Lebrun 1973), 73.5 (Backer and Bakhuizen van den Brink 1963-1968), and 74.3 (Borhidi 1991), respectively. Furthermore, the high numbers of alien species/log(area) for tropical islands suggest that sufficient numbers of suitable propagules of alien species have been successfully transported to both temperate and tropical areas (cf. Figs. 8.1B and 8.2B). Therefore, a natural resistance of the continental tropics to plant invasions clearly remains a possibility.

Fig 8.1. (page 156) Percentage of alien species (A), number of alien species per log(area) (B), and number of native species per log(area) (C) as a function of latitude in continental floras. In B the quadratic component of regressions on intervals between 50° N and 50° S are significant at 0.01 (S. and N. America) and 0.05 (Africa and Europe) probablility levels. In C the quadratic component of the regression for Americas is significant at 0.05 level, whereas the line for Africa and Europe is fitted by eye. Data on North American floras, including Mexico and Belize, are in Rejmánek and Randall (1994). Data on remaining areas, La Selva (Costa Rica), Panama, Chocó (Colombia), Guianas, Peru, São Paulo (Brazil), Buenos Aires (Argentina), Chile, Finland, Poland, The Netherlands, Germany, Portugal, Egypt, Senegal, Togo, Uganda, Rwanda, Selous (Tanzania), Namibia, Swaziland, South Africa, can be obtained by request from the author

This data set, however, does not suggest that the high diversity of native plant species is responsible for the resistance. First, in general, there are higher numbers of established alien plant species in the tropics of Central and South America than in areas of corresponding latitudes in Africa (Fig. 8.1B). The American tropics, nonetheless, exhibit higher species richness of native plants (Fig. 8.1C and Gentry 1988). Second, numbers of alien species/ log(area) on islands between 50° S and 50° N (Fig. 8.2B, n = 52) are independent of the absolute value of latitude (r=0.04, p > 0.7) but positively dependent on distance from the nearest continent (r=0.29,p < 0.05), and also positively dependent on the number of native species/log(Area) (r= 0.30, p < 0.05). It is, therefore, more likely that the isolation of islands with all the consequences (e.g., absence of large herbivores, low frequency of fires, interactions with and adaptation to only limited numbers of species through evolutionary time), rather than low species diversity per se, is responsible for their high invasibility (see also Carlquist 1974; Case 1981, Vitousek 1988; Pimm 1991; Merlin and Juvik 1992). The numbers of native species on islands seem to be an indicator of favorable conditions (actual evapo-transpiration? – see Wright 1983) for more invaders rather than an obstacle to invasions. Another part of the story could be the fact that wet and species-rich islands are also more suitable for agriculture and have been therefore more disturbed. We should be cautious, however, in making generalizations on the resistance of the tropics to plant invasions based just on numbers of species in local floras. It is well known that several African grasses are extremely successful in the American tropics (Daubenmire 1972; Parsons 1972). In some disturbed areas, their success is comparable only with the success of plants introduced from the Mediterranean Basin to California (Baker 1978; Rejmánek et al. 1991). One of the most invasive African grasses – *Melinis minutiflora* Beauv. – was originally described from Brazil! Total areas infested by alien plants would be certainly another measure of invasion vulnerability. Unfortunately, we do not have the data to make any rigorous comparisons.

Fig. 8.2. (page 159) Percentage of alien species (A), number of alien species per log(area) (B), and number of native species per log(area) (C) as a function of latitude in oceanic island floras. None of the variables exhibits any significant trend between 50° S and 50° N. Sixty-three islands from south to north: S. Shetland, Sandwich, Tierra del Fuego, Macquarie, Campbell, Falkland, Auckland, Marion and Prince Edward, New Zealand, Tristan and Cunha, Robben, Lord Howe, Kermadec, Henderson, Gambier, New Caledonia, Niuatoputapul, Madagascar, Rodrigues, Mauritius, Niue, Fiji, Aitutaki, St. Helena, Comoro, Christmas, Nukunkonu, Java, Seychelles, Canton, Galapagos, S. Tomé/Fernando Po/Príncipe/Annobon, Maldives, Ascension, Barbados, Wallis and Futuna, Guadaloupe, Guam, Gorda, Socorro, Puerto Rico, N. Marianas, Wake, Cayman, Cuba, Hawaii, Taiwan, Bahamas, Canary Islands, Tiburon, Guadalupe, Bermuda, Madeira, Santa Cruz, Wallops, Prince Edward, Newfoundland, Amchitka, Queen Charlotte, British Isles, Iceland, Greenland, Devon. Derived from various sources. Data on northern Atlantic and Caribbean islands are in Rejmánek and Randall (1994). Data on remaining islands can be obtained by request from the author.

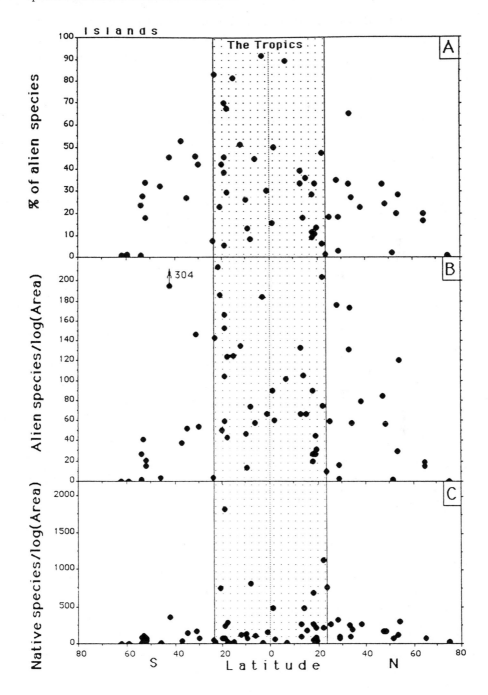

8.3 Intentional Introductions

Generalizations about biological invasions are mostly rather speculative because the numbers of failed introductions are usually not known (Simberloff
1989). Documented biological control studies, however biased they may be
(e.g., more carefully planned in developed countries), provide data about
failures. Of a total of 1054 fully documented introductions of insects as biological control agents of weeds (Julien 1992 and references therein), 378 (115
species) occurred in the tropics and 676 (176 species) in extratropical areas.
Analyses of these data led to the following conclusions: the overall rate of
establishment (permanent presence since introduction) was not significantly
different between tropical (56/90; 62.2%) and extratropical (403/632; 63.8%)
continental areas (G=0.018, $p>0.8$), or between tropical (188/288; 65.3%) and
extratropical (28/44; 63.6%) islands (G=0.01, $p>0.9$). Similarly, there was no
significant difference between relative numbers of species established in
tropical (26/31; 83.9%) and extratropical (127/162; 78.4%) continental areas
(G=0.054, $p>0.8$), and between number of species established on tropical
(69/115; 60.0%) and extratropical (26/35; 72.7%) islands (G=0.51, $p>0.4$). This
result is not entirely surprising because essentially all of the introductions of
biological control agents take place only when the host species (usually introduced weeds in depauperate communities) are abundant and widespread.
Data on biological control of insects are consistent with our findings: there
are no significant latitudinal differences in the proportion of successful biological control attempts (DeBach 1974; Price 1991).

Somewhat less rigorous data are available on bird introductions (Long
1981; Lever 1987). Nevertheless, if continental areas and islands >100 000 km^2
summarized by Long (1981; Table 8.1) are considered, the success of establishment seems to be higher in the tropics. Among extratropical areas, percentages of definitely established species in North America, Europe, Japan,
and New Zealand are 32.7, 39.1, 28.6, and 28.6, respectively. In completely
or prevailingly tropical areas, percentages of definitely established species
in southeast Asian mainland, South America, Africa-Arabia, Borneo, and
Cuba are 46.7, 50.0, 45.2, 57.1, and 30.0, respectively. The difference between means (32.3 and 45.8 %) is significant ($p < 0.05$) but additional factors should be taken into account. In general, because more species were
released in extratropical areas, lower rates of establishment may be expected (see Pimm 1991). On the other hand, unsuccessful introductions
were more often repeated in extratropical areas, increasing the chance of
final success. At any rate, it is conceivable that birds find fewer obstacles to
their establishment in the tropics than in extratropical areas. Constant food
availability and absence of severe winters may be important factors. The
absolute majority of permanent bird establishments takes place in disturbed landscapes. Apparently only very few introduced bird species have
penetrated into undisturbed tropical forests (see below).

Table 8.1. Woody species invading primary tropical forests
Code for areas of native range: Af=Africa, Am =South and Central America, As=Asia,
Au=Australia

Species (Family, life form, native to)		Introduced to:	reference
Albizia chinensis (Osbeck) Merr. (Leguminosae, tree, As)	G	East Usambara, Tanzania	6, 7
Ardisia crenata Sims (Myrsinaceae, shrub, As)	*	Mauritius	15
Ardisia humilis Vahl. (A. elliptica Thunb.? Myrsinaceae, shrub, As)	*	Rarotonga, Cook Islands	20
Caesalpinia pulcherrima (L.) Sw. (Leguminosae, shrub, Am)		Chamela, Jalisco, Mexico	12
Cinnamomum zeylanicum Blume (Lauraceae, small tree, As)	*	Seychelles	17, 18
Clidemia hirta (L.) D.Don. (Melastomataceae, shrub, Am)	*	Hawai'i; Seychelles East Usambara, Tanzania	7, 10, 17
Chrysophyllum cainito L. (Sapotaceae, tree, West Indies)	*	Mexico to northern South America	2
Cryptostegia grandiflora R. Br. (Asclepiadaceae, liana, India?)	G	Northern Australia	23
Cyphomandra betacea (Cavanilles) (Solanaceae, small tree, Am)	*	Kibale Forest, Uganda	5
Cyathea cooperi (F. Muell.) Domin (Cyatheaceae, small tree, Au)		Haleakala NP, Maui	22
Dioscorea alata L. (Dioscoreaceae, liana or vine, As)		East Usambara, Tanzania	7
Dioscorea bulbifera L. var. *antropophagorum* (A. Chev.) Summerth. (Dioscoreaceae; herb, liana or vine; As)		East Usambara, Tanzania	7
Dioscorea sansibarensis Pax (Dioscoreaceae; herb, liana or vine; Af)		Singapore	11
Duranta erecta L. (Verbenaceae, shrub, Am)		Queen Elizabeth NP, Uganda East Usambara, Tanzania	5, 7
Ficus altissima Blume (Moraceae, tree, As)	*	East Usambara, Tanzania	6, 7
Fuchsia arborescens Sims (Onagraceae, shrub, Am)		East Usambara, Tanzania	7
Hovenia dulcis Thunb. (Rhamnaceae, small tree, As)	*? G	East Usambara, Tanzania	7
Hura crepitans L. (Euphorbiaceae, tree, Am)	G	East Usambara, Tanzania	6
Ligustrum lucidum Ait. f. (Oleaceae, shrub, As)	*	Northern Argentina	16
Ligustrum robustum Blume (Oleaceae, shrub, As)	*	Mauritius; Réunion	4, 15
Litsea glutinosa C. B. Rob (Lauraceae, small tree, As)	*	Mauritius	15
Melastoma candidum D. Don (Melastomataceae, shrub, As)	*	Kaua'i and Hawai'i	10, 14

Table 8.1. cont.

Species (Family, life form, native to)		Introduced to:	reference
Maesopsis eminii Engl. (Rhamnaceae, tree, Af)	*G	East Usambara, Tanzania	1, 7
Medinilla venosa (Blume) Blume (Melastomataceae, shrub, As)	*	Maui, Hawai'i	10
Miconia calvescens DC (Melastomataceae, small tree, Am)		Tahiti; Maui (undisturbed forest?)	13, 10
Milletia dura Dunn (Fabaceae, tree, Af)	G	East Usambara, Tanzania	6, 7
Myrica faya Aiton (Myricaceae, small tree, Canary Islands)	*	Hawai'i	13
Ossaea marginata (Desr.) Triana (Melastomataceae, shrub, Am)	*	Mauritius	15
Passiflora edulis Sims (Passifloraceae, liana, Am)	*G	East Usambara, Tanzania	7
Passiflora molissima (Kunth) L.H. Bailey (Passifloraceae, liana, Am)	*	Hawai'i	10
Pittosporum undulatum Venten. (Pittosporaceae, shrub, Au)	*	Jamaica	8
Psidium cattleianum Sabine (Myrtaceae, small tree, Am)	*	Hawai'i; Mauritius; Réunion	4, 10, 15
Psidium guajava L. (Myrtaceae, small tree, Am)	*	Galápagos	21
Roystonea oleracea (Jacquin) Cook (Arecaceae, tree, Venezuela)	*?	Approuague, French Guiana	3
Rubus alceifolius Poiret (Rosaceae, shrub, As)	*	Réunion	4
Rubus ellipticus Sm. (Rosaceae, shrub, As)	*G	Hawai'i	13
Syzygium (*Eugenia*) *jambos* (L.) Alston (Myrtaceae, small tree, Am)	*	Puerto Rico; Hawai'i	9, 10
Syzigium (*Eugenia*) *malaccensis* (L.) Alston (Myrtaceae, small tree, As)	*	Iles Marquises	19
Spathodea campanulata P. Beauv. (Bignoniaceae, tree, Af)	G	East Usambara, Tanzania	7
Tetrazygia bicolor (Mill.) Cogn. (Melastomataceae, shrub, Am)	*	Nanawale Forest Reserve, Hawai'i	10
Thunbergia grandiflora Roxb. (Acanthaceae, liana, India)		Singapore; Northern Australia	11, 23
Wikstroemia indica (L.) C.A. Meyer (Thymelaeaceae, shrub, As)	*	Mauritius	15

*=Dispersed primarily by vertebrates, G reported only from gaps. Sources: 1, Binggeli and Hamilton (1993); 2, Croat (1978); 3, de Granville (1989); 4, MacDonald et al. (1991); 5, Rejmánek et al. (1996); 6, Rufo et al. (1989); 7, Sheil (1994); 8, Sugden et al. (1985); 9, Frank Wadsworth (pers. comm.) 10, Wester and Wood (1977); 11, Whitmore (1991); 12, Lott (1993); Alfredo Pérez Jiménez, pers. comm.; 13, Smith (1985); 14, Jacobi and Warshauer (1992); 15, Lorence and Sussman (1986); 16, E. Zardini, sec. Lorence and Sussman (1988); 17, Gerlach (1993); 18, Sauer (1988); 19, Halle (1978); 20, Merlin (1985); 21, Lawesson and Ortiz (1990); 22, Anderson et al. (1992); 23, Stella Humphries and Richard Groves (pers. comm.).

8.4 Invasions into Undisturbed Tropical Forests

Whitmore (1991) summarized information about 21 woody invasive species in perhumid tropical climates based mainly on his experience from southeast Asia. All except three, however, are species recorded from disturbed or human-made sites. Two of the exceptions are the climbers (*Dioscorea sansibarensis* and *Thunbergia grandiflora*), which are spreading in Singapore, the third is *Pittosporum undulatum*, which invades montane rain forests in Jamaica. Based on published records, personal communications, and my own field experience, I compiled a list of only 42 woody species (including Whitmore's three species) known to be invading at least one area of primary tropical forest (Table 8.1). A similar list for extratropical primary forests (not shown) is more than twice as long. Moreover, 21 species in Table 8.1 are known to be invasive in primary forests only on some islands, and out of the 21 species invading continental primary forests, eight are reported only from gaps (Table 8.1). In general, even island tropical forests seem to be resistant to plant invasisons as long as they are not disturbed by human activities or by introduced mammals (Watts 1970; Gerrish and Mueller-Dombois 1980; Parnell et al. 1989).

So far, the maximum number of alien woody species invading primary forests in a single continental area is reported from the East Usambara Mountains, eastern Tanzania: six species in 'intact' forests plus seven species invading gaps (Sheil 1994). The island-like character of forests in this area may be responsible for these relatively high numbers. Also, the East Usambaras have been a site for extensive crop and forestry trials since the end of the last century. Since analogical Atlantic rain forests of Brazil apparently have not been invaded to a comparable extent (Mori and Boom 1983; Por 1992; Grady Webster, pers. comm.), the second factor may be more important.

No one growth form dominates among successful woody invaders (Table 8.1): 19 trees (11 small trees), 16 shrubs, 7 lianas. Interestingly, however, at least 26 (62%) of the 42 species in Table 8.1 have fleshy fruits and are dispersed by mammals or birds. The hypothesis that possessing fleshy fruits may facilitate invasiveness is supported by the fact that herbaceous species, which mostly do not possess fleshy fruits, are even less successful in invading primary tropical forests than are woody species. The list of herbaceous invaders seems to be much shorter: *Adiantum raddianum* Presl (Adiantaceae; Africa; Shelpe 1970, Sheil 1994), *Blechnum occidentale* L. (Polypodiaceae; Hawai'i, Anderson et al. 1992), *Eleusine indica* (L.) Gaertn. (Poaceae; Jalisco, Mexico; McVaugh 1983, Alfredo Perez Jimenez, pers. comm.), *Mina lobata* Cerv. (Convolvulaceae; E. Usumbara, Tanzania; Sheil 1994), *Olyra latifolia* L. (Poaceae; several Forest Reserves in Uganda; Rejmánek et al. 1995), *Oplismenus compositus* (L.) P. Beauv. and *O. hirtellus* (L.) P. Beauv. (Poaceae; Hawai'i; Wagner et al. 1990), *Paspalum conjugatum*

Bergius (Poaceae; Maui; Anderson et al. 1992), *Pteris cretica* L. and *P. tripartita* Sw. (Pteridaceae; Peru and very likely forests in other areas in tropical America; Tryon and Stolze 1989), *Stachytarpheta urticifolia* Sims. (Verbenaceae; E. Usumbara, Tanzania; Sheil 1994), *Thelypteris opulenta* (Kaulf.) Fosberg (=*T. extensa* (Blume) Mort., Thelypteridaceae; Canal Zone, Panama, Croat 1978; Peru, Smith 1992), *Vanilla fragrans* (Salisb.) Ames (Orchideaceae; Barro Colorado Island; Croat 1978). One or two herbaceous invaders in the genus *Hedychyum* (Zingiberaceae) are dispersed in primary forests by birds because of their fleshy arils (Hawai'i, Smith 1985; Réunion, Macdonald et al.1991).

There are a few plant species widespread in undisturbed tropical forests on more than one continent. For example, *Carapa procera* DC., *Geophyla repens* (L.) I.M. Johnston, and *Parinari excelsa* Sabine are components of both African and South American forests. Such species represent a special challenge for future research. Some of these species grow also along streams close to the coast (*Paullinia pinata* L., *Symphonia globulifera* L.) and therefore their intercontinental dispersal by sea currents is possible (Thorne 1973; Gunn and Dennis 1976).

The fact that very few plant species are invading 'undisturbed' primary tropical forests does not mean that exotic plants do not influence regeneration of disturbed forests. In Uganda, for example, populations of *Acacia mearnsii* De Wild., *Broussonietia papyrifera* Vent., *Senna (Cassia) siamea* (Lam.) Irwin & Barneby, *S. spectabilis* (DC.) Irwin & Barneby, and *Solanum mauritianum* Scopoli represent serious obstacles for postagricultural forest recovery in several Forest Reserves and National Parks (Rejmánek et al. 1996). Even under continuous import of mature forest propagules, secondary succession is blocked by these species for at least 15 years. Similarly, *Austroeupatorium (Eupatorium) inulaefolium* (H.B.K.) King & Robinson, *Cecropia peltata* L., *Chromolaena odorata* (L.) King & Robinson, *Cinchona succirubra* Pavon ex Klotzsch, *Eugenia cumini* (L.) Druce, *Fraxinus uhdei* (Wenzig) Lingelsheim, *Lantana camara* L., *Leucaena leucocephala* (Lam.) de Wit, *Mimosa pigra* L., *Rubus argutus* Link, *R. moluccanus* L. and many others interfere negatively with regeneration of tropical forests in many countries where these species were introduced (Koechlin et al. 1974; Guillaumet 1984; Smith 1985; Macdonald *et al.* 1988, McKey 1988; de Rouw 1991; Whitmore 1991; Adam 1992; Smiet 1992). Besides competing with seedlings of native species, several exotic grasses are also responsible for higher frequency of fires and subsequent elimination of native seedlings (D'Antonio and Vitousek 1992).

There are very few documented cases of insect invasions into primary tropical forests (Bock 1980; Roubik 1988; S. J. Wright, Chap. 2, this Vol.). Also, very few non-native vertebrates are invading primary tropical forests, and these are mainly on islands (Roots 1976; Lever 1985, 1987; Sussman and Tattersall 1986; Savidge 1987; Emmons 1990; Hutterer and Tranier 1990, Mackinnon and Phillipps 1993; Harrison, 1968 in Primack 1993; Sick

1993). Out of about 45 exotic bird species established in Hawai'i, only nine have invaded native forests: *Cettia diphone* (Japanese bush-warbler), *Copsychus malabricus* (white-rumped shama), *Garrulax pectoralis* (greater necklaced laughing-thrush), *G. caerulatus* (Gray-sided laughing-thrush), *G. canorus* (Hwamei), *Leiothrix lutea* (red-billed leiothrix), *Zosterops japonicus* (Japanese white-eye), *Cardinalis cardinalis* (northern cardinal), *Streptopelia chinensis* (spotted dove) (Hawai'i Audubon Society 1993; Williamson and Fitter 1996).

However, several of the alien vertebrates established in tropical forests are more influential than any of the alien plants: water buffalo (*Bubalus bubalis*), pig (*Sus scrofa*), mongoose (*Herpestes auropunctatus*), feral cat (*Felis catus*), brown tree snake (*Boiga irregularis*) (Stone 1985; Savidge 1991; Russell-Smith and Bowman 1992). Some introduced mammals are extremely important in creating disturbance and providing dispersal for alien plant species in tropical forests (Smith 1985, 1992; Stone 1985; Sussman and Tattersall 1986).

8.5 Speculations

Everything else being equal, the higher resistance of continental tropical communities, and forests in particular, to invasions of plants (and probably some other groups of biota) may be due to both biotic and abiotic factors. Evidence that high species diversity of native flora is directly responsible for higher resistance to invasions is weak. In Semliki Forest Reserve, Uganda, for example, we failed to find any difference in invasibility between species-rich communities and *Cynometra alexandri* monodominant forests (Rejmánek et al. 1996). Incidentally, the most important extratropical centers of plant species diversity - the Cape Floral Region and southern Australia - are notoriously vulnerable to plant invasions (Kloot 1991; Wells 1991).

Available evidence suggests that the high rate of recovery of wet tropical vegetation after natural and human-made disturbances is an important part of the story. This high rate of recovery has its abiotic (high moisture, radiation, temperature) and biotic (pool of immediately germinating, fast growing native and, very often, some already naturalized alien species) components. Very few invaders can survive in the presence of such robust and fast-growing pioneer species as *Cecropia* spp., *Cyperus papyrus, Gynerium sagittatum, Lantana camara, Panicum maximum, Penisetum purpureum, Saccharum spontaneum, Sesbania emerus, Tessaria integrifolia,* or *Trema* spp. It is conceivable that the scarcity of such species on islands is partly responsible for their vulnerability to plant invasions. Loss of dispersibility in island plants (Carlquist 1974) may contribute to slower vegetation recovery on islands as well. High species diversity can contribute to the biotic component of vegetation recovery; McNaughton (1985) and Tilman and Downing (1994) found a positive correlation between plant biomass recovery and species diversity in their studies of tropical and temperate grasslands.

In contrast to plants, introduced birds may be favored to some extent in the tropics (mainly in disturbed areas) due to their mobility and the presence of sufficient food supplies year-round.

Mature tropical rain forests can be invaded by only very few (mostly shade-tolerant) plant species. More than 60% of woody species invading primary tropical rain forests are dispersed by native or introduced vertebrates. This is consistent with the fact that usually more than 60% of woody species in wet tropical forests are adapted for seed dispersal by vertebrates (Howe and Smallwood 1982; Gentry 1983; Willson et al. 1989; Levey et al. 1994). The same vertebrate species that are responsible for the maintenance of diversity in tropical forests may also disperse the seeds of exotic species. Fates of seeds of species with fleshy fruits, however, are dependent on many factors other than the presence of potential dispersers, which may be why even shade-tolerant woody species with fleshy fruits are rather rarely invading tropical forests. Total fruit abundance, social and foraging behavior, and insect seed predation may be involved (Chapman 1989; Willson and Crome 1989; Gautier-Hion et al. 1993).

Whether primary tropical dry forests are less or equally resistant to invasions than primary tropical rain forests remains an open question. Tropical deciduous and semideciduous forests in Mexico do not seem to be more vulnerable to invasions than rain forests (Lott 1993; Alfredo Pérez Jiménez, pers. comm.). Nevertheless, remnants of tropical dry forests probably will be invaded by larger numbers of non-native plant species because they have been so highly disturbed by people (Murphy and Lugo 1986; Sussman and Rakotozafy 1994). In general, there are good theoretical (Kindlmann and Rejmánek 1982; Shmida and Ellner 1984) and empirical (Janzen 1983; Hamilton and Bensted-Smith 1989) reasons to expect that fragmented tropical forests are more vulnerable to invasions than continuous forest areas.

Acknowledgments. This chapter benefited from discussions and correspondence with Pierre Binggeli, Rodolfo Dirzo, Anthony Katende, Alfredo Pérez Jiménez, Stuart L. Pimm, John Randall, Eliska Rejmánková, Frank Wadsworth, Grady L. Webster, Mark Williamson, and all the participants of the Workshop on Ecosystem Functions of Biodiversity in Tropical Forests. The research was partially supported by the National Geographic Society.

References

Adam P (1992) Australian rainforests. Clarendon Press, Oxford

Allee WC, Emerson AE, Park O, Park T, Schmidt KP (1949) Principles of animal ecology. Saunders, Philadelphia

Anderson SJ, Stone CP, Higashino PK (1992) Distribution and spread of alien plants in Kipahulu Valley, Haleakala National Park. In: Stone CP, Smith CW, Tunison, JT (eds) Alien plant invasions in native ecosystems of Hawai'i: management and research. Univ Hawai'i, Honolulu, pp 300-338

Backer CA, Bakhuizen van den Brink RC (1963-1968) Flora of Java, 3 vols. Wolters-Noordhoff, Groningen

Baker HG (1978) Invasion and replacement in Californian and neotropical grasslands. In: Wilson JR (ed) Plant relations and pastures. CSIRO, Melbourne, pp 368-384

Binggeli P, Hamilton AC (1993) Biological invasion by *Maesopsis eminii* in the East Usambara forest, Tanzania. Opera Bot 121:229-235.

Bock IR (1980) Current status of the *Drosophila melanogaster* species group. Syst Entomol 5:341-356

Borhidi A (1991) Phytogeography and vegetation ecology of Cuba. Akademiai Kiado, Budapest

Brown BJ, Ewel JJ (1988) Responses to defoliation of species-rich and monospecific tropical plant communities. Oecologia 75:12-19

Chapman CA (1989) Primate seed dispersal: the fate of dispersed seeds. Biotropica 21:148-154

Carlquist S (1974) Island biology. Columbia Univ Press, New York

Case TJ (1981) Niche packing and coevolution in competition communities. Proc Natl Acad Sci USA 78: 5021-5025.

Case TJ (1990) Invasion resistance arises in strongly interacting species-rich model competition communities. Proc Natl Acad Sci USA 87:9610-9614

Case TJ (1991) Invasion resistance, species build-up and community collapse in metapopulation models with interspecific competition. Biol J Linn Soc 42:239-266

Clarkson JR, Kenneally KF (1988) The floras of Cape York and the Kimberley: a preliminary comparative analysis. Proc Ecol Soc Aust 15:259-266

Cowie ID, Werner PA (1993) Alien plant species invasive in Kakadu National Park, tropical northern Australia. Biol Conserv 63:1278- 135

Croat TB (1978) Flora of Barro Colorado Island. Stanford Univ Press, Stanford

D'Antonio CM, Vitousek PM (1992) Biological invasions by exotic grasses, the grass/fire cycle, and global change. Annu Rev Ecol Syst 23:63-87

Daubenmire R (1972) Ecology of *Hyparrhenia rufa* (Nees) Stapf in derived savanna in north-western Costa Rica. J Appl Ecol 9:11-23

DeBach P (1974) Successes, trends, and future possibilities. In: DeBach P (ed) Biological control of insect pests and weeds. Reinhold, New York, pp 673-713

De Granville JJ (1989) La distribución de las palmas en la Guyana Francesa. Acta Amazon 19:115-138

De Rouw A (1991) The invasion of *Chromolaena odorata* (L.) King & Robinson (ex *Eupatorium odoratum*), and competition with the native flora, in rain forest zone, south-west Cote d'Ivoire. J Biogeogr 18:13-23

Drake JA (1988) Models of community assembly and the structure of ecological landscapes. In: Gross L, Hallam T, Levin S (eds) Proceedings of the International Conference on Mathematical Ecology. World Press, Singapore, pp 585-604

Emmons LH (1990) Neotropical rainforest mammals. Univ Chicago Press, Chicago

Elton CS (1958) The ecology of invasions by animals and plants. Methuen, London

Farnworth EG, Golley FB (eds) (1974) Fragile ecosystems. Springer, Berlin Heidelberg New York

Futuyma DJ (1973) Community structure and stability in constant environments. Am Nat 107:443-446

Gautier-Hion A, Gautier JP, Maisels F (1993) Seed dispersal versus seed predation: an intersite comparison of two related African monkeys. Vegetatio 107/108:237-244

Gerrish G, Mueller-Dombois D (1980) Behavior of native and non-native plants in two tropical rain forests on Oahu, Hawaiian Islands. Phytocoenologia 8:237-295

Gentry AH (1983) Dispersal ecology and diversity in Neotropical forest communities. Sonderb Naturwiss Ver Hamb 7:303-314

Gentry AH (1988) Changes in plant community diversity and floristic composition on environmental and geographical gradients. Ann Mo Bot Gard 75:1-34

Gerlach J (1993) Invasive Melastomataceae in Seychelles. Oryx 27:22-26

Guillaumet JL (1984) The vegetation: an extraordinary diversity. In: Jolly A, Oberlé P, Albignac R (eds) Madagascar. Pergamon Press, Oxford, pp 27-54

Gunn CR, Dennis JV (1976) World guide to tropical drift seeds and fruits. Quadrangle/The New York Times Book C, New York

Halle F (1978) Arbres et forêts des Iles Marquises. Cah Pac 21:315-357

Hallett JG (1991) Structure and stability of small mammal faunas. Oecologia 88:383-393

Hamilton AC, Bensted-Smith R (eds) (1989) Forest conservation in the East Usambara Mountains, Tanzania. The IUCN Trop For Progr, IUCN, Gland, Switzerland

Hannah L, Lohse D, Hutchinson C, Carr JL, Lankerani A (1994) A preliminary inventory of human disturbance of world ecosystems. Ambio 23:246-250

Harrison GW (1979) Stability under environmental stress: resistance, resilience, persistence, and variability. Am Nat 113:659-669

Hawai'i Audubon Society (1993) Hawai'i's birds, 4th edn. Hawai'i Adubon Soc, Honolulu

Hobbs RJ (1992) Is biodiversity important for ecosystem functioning? In: Hobbs RJ (ed) Biodiversity of mediterranean ecosystems in Australia. Surrey Beatty & Sons, Chipping Norton, pp 211-229

Hobbs RJ, Huenneke LF (1992) Disturbance, diversity, and invasion: implications for conservation. Conserv Biol 6: 324-337.

Howe HF, Smallwood J. (1982) Ecology of seed dispersal. Annu Rev Ecol Syst 13:201-228

Hundley HG, Chit KK (1961) List of trees, shrubs, herbs and principal climbers recorded from Burma, 3rd edn. Government Printing Press, Rangoon

Hutterer R, Tranier M (1990) The immigration of the Asian house shrew (*Suncus murinus*) into Africa and Madagascar. In: Peters G, Hutterer R (eds) Vertebrates in the tropics. Museum Alexander Koenig, Bonn, pp 211-229

Jacobi JD, Warshauer FR (1992) Distribution of six alien plant species in upland habitats on the Island of Hawai'i. In: Stone CP, Smith CW, Tunison, JT (eds) Alien plant invasions in native ecosystems of Hawai'i: management and research. Univ Hawai'i, Honolulu, pp 155-179

Janzen DH (1983) No park is an island: increase in interference from outside as park size decreases. Oikos 41:402-410

Julien MH (ed) (1992) Biological control of weeds. A world catalogue of agents and their target weeds. CAB International, Oxon, UK

Kindlmann P, Rejmánek M (1982) Number of species at stable equilibrium of complex model ecosystems. Ecol Model 16:85-90

Kloot PM (1991) Invasive plants of southern Australia. In: Groves RH, Di Castri F (eds) Biogeography of mediterranean invasions. Cambridge Univ Press, Cambridge, pp 131-143

Koechlin J, Guillaumet JL, Morat P (1974) Flore et Végétation de Madagascar. Cramer, Vaduz

Lawesson JE, Ortiz L (1990) Plantas introducidas en las Islas Galápagos. Monogr Syst Bot Mo Bot Gard 32:201-210

Lebrun JP (1973) Enumération des plantes vasculaires du Sénégal. Institut d'Elevage et de Médicine Vétérinaire des Pays Tropicaux, Maison-Alfort

Leigh EG (1975) Population fluctuations, community stability, and environmental variability. In: Cody ML, Diamond JM (eds) Ecology and evolution of communities. Harvard Univ Press, Cambridge, Mass, pp 51-73

Lever C (1985) Naturalized mammals of the world. Longman, London

Lever C (1987) Naturalized birds of the world. Longman, London

Levey DJ, Moermond TC, Denslow JS (1994) Frugivory: an overview. In: McDade LA, Bawa KS, Hespenheide HA, Hartshorn GS (eds) La Selva. Ecology and natural history of a Neotropical rain forest. Univ Chicago Press, Chicago, pp 282-294

Long JL (1981) Introduced birds of the world. David & Charles, London

Lorence DH, Sussman RW (1986) Exotic species invasion into Mauritius wet forest remnants. J Trop Ecol 2:147-162

Lorence DH, Sussman RW (1988) Diversity, density, and invasion in a Mauritian wet forest. Monogr Syst Bot Mo Bot Gard 25:187-204

Lott EJ (1993) Annotated checklist of the vascular flora of the Chamela Bay region, Jalisco, Mexico. Occas Pap Calif Acad Sci 148:1-60

MacArthur RH (1970) Species packing and competitive equilibrium for many species. Theor Popul Biol 1:1-11

Macdonald IAW, Ortiz L, Lawesson JE, Nowak JB (1988) The invasion of highlands in Galápagos by Red Quinine-tree *Cinchona succirubra*. Environ Conserv 15:215-220

Macdonald IAW, Thébaud C, Strahm WA (1991) Effects of alien plant invasions on native vegetation remnants on La Réunion (Mascarene Islands, Indian Ocean). Environ Conserv 18:51-61

Mackinnon J, Phillipps K (1993) The birds of Borneo, Sumatra, Java, and Bali. Oxford Univ Press, Oxford

Maury-Lechon G, Hadley M, Younes Y (eds) (1984) The significance of species diversity in tropical forest ecosystems. Biol Int Spec Issue 6, Int Union Biol Sci, Paris

May RM (1973) Stability and complexity in model ecosystems. Princeton Univ Press, Princeton

McKey D (1988) *Cecropia peltata*, an introduced Neotropical pioneer tree, is replacing *Musanga cecropioides* in southwestern Cameroon. Biotropica 20:262-264

McNaughton SJ (1978) Stability and diversity of ecological communities. Nature 274:251-253

McNaughton SJ (1985) Ecology of a grazing ecosystem: the Serengeti. Ecol Monogr 55:259-294

McNaughton SJ (1993) Biodiversity and function of grazing ecosystems. In: Schulze E-D, Mooney HA (eds) Biodiversity and ecosystem function. Springer, Berlin Heidelberg New York, pp 361-383

McVaugh R (1983) Gramineae, vol 14. Flora Novo-Galiciana. Univ Michigan Press, Ann Arbor

Merlin MD (1985) Woody vegetation in the upland region of Rarotonga, Cook Islands. Pac Sci 39:81-99

Merlin MD, Juvik JO (1992) Relationships among native and alien plants on Pacific islands with and without significant human disturbance and feral ungulates. In: Stone CP, Smith CW, Tunison, JT (eds) Alien plant invasions in native ecosystems of Hawai'i: management and research. Univ Hawai'i, Honolulu, pp 597-624

Mori SA, Boom BM (1983) Southern Bahian moist forests. Bot Rev 49:155-232

Murphy PG, Lugo AE (1986) Ecology of tropical dry forest. Annu Rev Ecol Syst 17:67-88

Naeem S, Thompson LJ, Lawler SP, Lawton JH, Woodfin RM (1994) Declining biodiversity can alter the performance of ecosystems. Nature 368:734-737

Orians GH (1975) Diversity, stability and maturity in natural ecosystems. In: van Dobben WH, Lowe-McConnell RH (eds) Unifying concepts in ecology. Dr W Junk, The Hague, pp 139-150

Parnell JAN, Cronk Q, Jackson PW, Strahm W (1989) A study of the ecological history, vegetation and conservation management of Ile aux Aigrettes, Mauritius. J Trop Ecol 5:355-374

Parsons J (1972) Spread of African pasture grasses to the American tropics. J Range Manage 25:12-17

Pimm SL (1984) The complexity and stability of ecosystems. Nature 307:321-326

Pimm SL (1991) Balance of nature? Univ Chicago Press, Chicago

Pimm SL (1993) Biodiversity and the balance of nature. In: Schulze E-D, Mooney HA (eds) Biodiversity and ecosystem function. Springer, Berlin Heidelberg New York, pp 347-359

Por FD (1992) Sooretama the Atlantic rain forest of Brazil. SPB Acad Publ, The Hague

Post WM, Pimm SL (1983) Community assembly and food web stability. Math Biosci 64:169-192

Price PW (1991) Patterns in communities along latitudinal gradients. In: Price PW, Lewinsohn TM, Fernandes GW, Benson WW (eds) Plant-animal interactions. Evolutionary ecology in tropical and temperate regions. Wiley, New York, pp 51-70

Primack RB (1993) Essentials of conservation biology. Sinauer, Sunderland, MA

Rejmánek M (1989) Invasibility of plant communities. In: Drake JA, Mooney HA, di Castri F, Groves RH, Kruger FJ, Rejmanek M, Williamson M (eds) Biological invasions: a global perspective. Wiley, Chichester, pp 369-388

Rejmánek M (1992) Stability in a multi-species assemblage of large herbivores in East Africa: an alternative interpretation. Oecologia 89:454-456

Rejmánek M, Stary P (1979) Connectance in real biotic communities and critical values for stability of model ecosystems. Nature 280: 311-313.

Rejmánek M, Randall JM (1994) Invasive alien plants in California: 1993 summary and comparison with other areas in North America. Madroño 41:161-177

Rejmánek M, Thomsen CD, Peters ID (1991) Invasive vascular plants of California. In: Groves RH, Di Castri F (eds) Biogeography of mediterranean invasions. Cambridge Univ Press, Cambridge, pp 81-101

Rejmánek M, Rejmánková E, Katende A (1996) Invasive plant species in Ugandan protected areas. Res Explor (in press)

Robinson JV, Valentine WD (1979) The concepts of elasticity, invulnerability, and invadability. J Theor Biol 81:91-104

Roubik DW (1988) Ecology and natural history of tropical bees. Cambridge Univ Press, Cambridge

Roxburgh SH (1994) Estimation of the parameters of the community matrix and implications for community stability using a lawn community. PhD Thesis, Univ Otago, Dunedin, New Zealand

Roots C (1976) Animal invaders. Universe Books, New York

Rufo CK, Mmari C, Kibuwa SP, Lowett J, Iverson S, Hamilton AC (1989) A preliminary list of plant species recorded from the East Usambara Forests. In: Hamilton AC, Bensted-Smith R (eds) Forest conservation in the East Usambara Mountains, Tanzania. IUCN, Gland, Swizterland, pp 157-179

Russell-Smith J, Bowman DMJS (1992) Conservation of monsoon rainforest isolates in the Northern Territory, Australia. Biol Conserv 59:51-64

Sauer JD (1988) Plant migration. Univ California Press, Berkeley

Savidge JA (1987) Extinction of an island forest avifauna by an introduced snake. Ecology 68:660-668

Savidge JA (1991) Population characteristics of the introduced brown tree snake (*Boiga irregularis*) on Guam. Biotropica 23:294-300

Seifert RP, Seifert FH (1976) A community matrix analysis of *Heliconia* insect communities. Am Nat 110:461-483

Sheil D (1994) Naturalized and invasive plant species in the evergreen forests of the East Usambara Mountains, Tanzania. Afr J Ecol 32:66-71

Shelpe EACLE (1970) Pteridophyta. Flora Zambesiaca. Crown Agents for Oversea Governments, London

Shmida A, Ellner S (1984) Coexistence of plant species with similar niches. Vegetatio 58:29-55

Sick H (1993) Birds in Brazil. Princeton Univ Press, Princeton

Simberloff D (1989) Which insect introductions succeed and which fail? In: Drake JA, Mooney HA, di Castri F, Groves RH, Kruger FJ, Rejmanek M, Williamson M (eds) Biological invasions: a global perspective. Wiley, Chichester, pp 61-75

Smiet AC (1992) Forest ecology on Java: human impact and vegetation of montane forest. J Trop Ecol 8:129-152

Smith AR (1992) Thelypteridaceae. Pteridophyta of Peru. Part III/16. Fieldiana Bot New Ser 29: 1-80

Smith CW (1985) Impact of alien plants on Hawai'i's native biota. In: Stone CP, Scott JM (eds) Hawai'i's terrestrial ecosystems, preservation and management. Univ Hawai'i, Honolulu, pp 180-250

Smith CW (1992) Distribution, status, phenology, rate of spread, and management of Clidemia in Hawai'i. In: Stone CP, Smith CW, Tunison JT (eds) Alien plant invasions in native ecosystems of Hawai'i: management and research. Univ Hawai'i, Honolulu, pp 242-253

Springett JA (1976) The effect of planting Pinus pinaster Ait. on populations of soil micro-arthropods and on litter decomposition at Gnangara, Western Australia. Aust J Ecol 1:83-87

Stone CP (1985) Alien animals in Hawai'i's native ecosystems: toward controlling the adverse effects of introduced vertebrates. In: Stone CP, Scott JM (eds) Hawai'i's terrestrial ecosystems, preservation and management. Univ Hawai'i, Honolulu, pp 251-297

Sugden AM, Tanner EVJ, Kapos V (1985) Regeneration following clearing in a Jamaican montane forest: results of a ten-year study. J Trop Ecol 1:329-351

Sussman RW, Rakotozafy A (1994) Plant diversity and structural analysis of a tropical dry forest in southwestern Madagascar. Biotropica 26:241-254

Sussman RW, Tattersall I (1986) Distribution, abundance, and putative ecological strategy of Macaca fascicularis on the Island of Mauritius, Southwestern Indian Ocean. Folia Primatol 46:28-43

Thorne RF (1973) Floristic relationships between tropical Africa and tropical America. In: Meggers, BJ, Ayensu ES, Duckworth WD (eds) Tropical forest ecosystems in Africa and South America: a comparative review. Smithsonian Inst Press, Washington DC, pp 27-47

Tilman D, Downing JA (1994) Biodiversity and stability in grasslands. Nature 367:363-365

Trepl L (1990) Zum Problem der Resistenz von Pflanzengesellschaften gegen biologische Invasionen. Verh Berl Bot Ver 8:195-230

Tryon RM, Stolze RG (1989) Pteridophyta of Peru. Part II. Fieldiana Bot New Ser 22:1-128

Usher MB (1991) Biological invasions into tropical nature reserves. In: Ramakrishnan PS (ed) Ecology of biological invasion in the tropics. Int Sci Publ, New Delhi, pp 21-34

Vitousek PM (1988) Diversity and biological invasions of oceanic islands. In: Wilson EO (ed) Biodiversity. Nat Acad Press, Washington, pp 181-189

Vitousek PM, Hooper DU (1993) Biological diversity and terrestrial ecosystem biogeo-chemistry. In: Schulze E-D, Mooney HA (eds) Biodiversity and ecosystem function. Springer, Berlin Heidelberg New York, pp 3-14

Wagner WL, Herbst DR, Sohmer SH (1990) Manual of the flowering plants of Hawai'i, vol. 2. Univ Hawaii Press, Honolulu

Watts D (1970) Persistence and change in the vegetation of oceanic islands: an example from Barbados, West Indies. Can Geogr 14:91-109

Wells MJ (1991) Introduced plants of the fynbos biome of South Africa. In: Groves RH, Di Castri F (eds) Biogeography of mediterranean invasions. Cambridge Univ Press, Cambridge, pp 115-129

Wester LL, Wood HB (1977) Koster's curse (Clidemia hirta), a weed pest in Hawaiian forests. Envir Conserv 4:35-41

Whitmore TC (1991) Invasive woody plants in perhumid tropical climates. In: Ramakrishnan PS (ed) Ecology of biological invasion in the tropics. Int Sci Publ, New Delhi, pp 35-40

Williamson M, Fitter A (1996) The varying success of invaders. Ecology

Willson MF, Crome FHJ (1989) Patterns of seed rain at the edge of a tropical Queensland rain forest. J Trop Ecol 5:301-308

Willson MF, Irvine AK, Walsh NG (1989) Vertebrate dispersal syndromes in some Australian and New Zealand plant communities, with geographic comparisons. Biotropica 21:133-147

Wolda H (1978) Fluctuations in abundance of tropical insects. Am Nat 112:1017-1045
Wolda H (1983) Long-term stability of tropical insect populations. Res Popul Ecol, Suppl 3:112-126
Wright DH (1983) Species-energy theory: an extension of species-area theory. Oikos 41:496-506

9 The Role of Biodiversity in Tropical Managed Ecosystems

Alison G. Power[1] and Alexander S. Flecker[1,2]

9.1 Introduction

Efforts to preserve biodiversity have been focused primarily on remaining areas of natural ecosystems, but only 5% of the terrestrial environment is unmanaged and uninhabited (Western and Pearl 1989), and only 3.2% is protected in national parks (Reid and Miller 1989). Of the 95% of the world's land devoted to managed ecosystems, approximately 50% is in agriculture, 20% in commercial forestry, and 25% in human settlements, such as cities, towns, and villages (Western and Pearl 1989). Although the biodiversity of any particular managed ecosystem may be low, a large proportion of the total species of a region may live in such systems (Pimentel et al. 1992). The extent to which biodiversity and ecosystem processes are modified by management varies tremendously. These data suggest that more attention should be paid to understanding patterns of biodiversity in managed ecosystems and how species richness influences the functioning of those systems.

In most managed ecosystems in tropical regions, the major management goal is high primary productivity. Because plant diversity is purposely varied to achieve this goal, managed systems represent convenient opportunities for testing hypotheses about relationships between species diversity and ecosystem functioning. Unfortunately, relatively few studies have been designed with this aim explicitly in mind. A further constraint in trying to generalize from insights derived from the study of managed ecosystems is that many of the species, including managed plants, their pests, and natural enemies, are aliens that lack a significant evolutionary history to the site of study. Nevertheless, examining the linkages between biodiversity and ecosystem processes in managed ecosystems may be useful for developing better hypotheses about these linkages in less disturbed systems and eventually for designing more effective managed ecosystems.

---

[1] Section of Ecology and Systematics, Cornell University Ithaca, New York 14853, USA
[2] Center for the Environment, Cornell University Ithaca, New York 14853, USA

Ecological Studies, Vol. 122
Orians, Dirzo and Cushman (eds) Biodiversity and Ecosystem Processes in Tropical Forests
© Springer-Verlag Berlin Heidelberg 1996

In this chapter we focus on biodiversity and functioning in ecosystems managed for achieving high rates of plant production. We begin by describing briefly a number of major types of managed ecosystems in tropical forest regions and patterns of biodiversity in them, in order to illustrate the broad range of potential alterations in diversity and functioning of "tropical managed ecosystems". We then examine whether differences in plant diversity influence attributes of ecosystem functioning, focusing on primary productivity and rates of herbivory. Because the literature linking biodiversity and ecosystem processes in tropical managed ecosystems is sparse, our conclusions are necessarily speculative.

9.2 Examples of Tropical Managed Ecosystems

9.2.1 Managed Forests

Increasing evidence shows that forested areas long considered "pristine" have been modified by people for centuries (Denevan et al. 1984). Indigenous people altered the diversity and abundance of plant and animal species by managing forests for food, fiber, wood, medicines, and fuel. For example, Maya civilizations modified extensive areas of Central American tropical forest still thought of as primary forest (Barrera et al. 1977; Rico-Gray et al. 1985; Gómez-Pompa 1987). These managed forests or "forest gardens" mimic the physiognomy and diversity of the "natural" forests in the same region and are often virtually indistinguishable from them (Gordon 1982; Gómez-Pompa et al. 1987). "Human-made" or "artificial" rain forests are common in many parts of the tropics, including Central America (Wilken 1977), South America (Pinkley 1973; May 1984; Posey 1985), the Caribbean (Kimber 1966), and southeast Asia (Wiersum 1982; Torquebiau 1984). These systems are altered by the planting and maintenance of useful native species, the selective removal of undesirable species, and minimal disturbance of the understory (Alcorn 1990). Over the long-term, these practices produce a heterogenous association of useful native tree species. For example, the managed forests of the Huastec Maya of northeastern Mexico contain more than 300 species of plants, including 33 species used for construction material, 81 used for fruits or other edible parts, and more than 200 used for medicines (Alcorn 1983).

9.2.2 Home Gardens

Highly managed, mixed gardens of trees, shrubs, and herbaceous species close to a home are known as home gardens, kitchen gardens, or dooryard gardens. These systems typically are 0.5 to 2 ha in size (Gliessman 1990a), and, like forest gardens, they provide food, fodder, firewood, medicines, ornamental plants, and construction materials. They also often include domestic animals. Species richness in these systems ranges from relatively modest levels to the high species richness exhibited by managed forests. Allison (1983) found an average of 33 plant species in upland and 55 species in lowland home gardens in Mexico. A single Costa Rican home garden contained 71 plant species one year and 83 species the next in an area of 1240 m^2 (Gliessman 1990b). In a single community on the Yucatan Peninsula of Mexico, Herrera Castro (Herrera-Castro 1991) found 404 plant species in home gardens ranging in size from 600 to 6000 m^2. Surveys in Indonesia document the use of more than 600 plant species in home gardens (Brownrigg 1985), although no individual garden would contain this many species. Home garden systems are similar to forest gardens and managed forests, but management is typically much more intensive, species richness is often somewhat lower, and productivity is probably higher.

9.2.3 Swidden Agriculture

Another traditional use of tropical forests by indigenous populations is swidden cultivation (variously known as shifting cultivation or slash-and-burn agriculture). In shifting cultivation systems, temporary forest clearings are planted for a few years with annual or short-lived annual crops and then allowed to remain fallow longer than the cultivation period (NRC 1993). Direct comparisons of the species richness of swiddens and forest are rare, but the few relevant studies indicate strikingly lower species richness in the swiddens compared to the forest. Plant diversity in swidden plots of Jivaroan communities in the Peruvian Amazon generally contained fewer than 15 species, whereas forest plots contained several times that number (Boster 1983). Moreover, swidden plots were heavily dominated by manioc (*Manihot esculenta*) which comprised 81-83% of the individual plants in a swidden, but many different cultivars of manioc are typically grown in each plot. In contrast, the most abundant species encountered in forest plots never exceeded about 20% of the individuals. Despite low species richness in the early stages of swidden, the swidden system as a whole may be quite diverse over the entire cycle. For example, a study of the *kebun-talun* system of Java found 112 plant species in this swidden system compared to 127 species in home gardens (Christanty et al. 1986). Like many swidden systems, *kebun-talun* includes a long period of perennial production in a managed fallow, and species richness increases dramatically from the initial stage dominated by annual crops.

9.2.4 Intensive Annual and Perennial Crops

Although monocultures are not as dominant in tropical agricultural systems as they are in temperate ones, the number of plant species is low in many agroecosystems in the tropics. Much of the highly productive land in the humid tropics has been converted to intensive annual monocultures of both export crops and food crops, such as rice and maize (NRC, 1993). During the 1950s, huge areas of the Pacific lowlands of Central America were cleared of deciduous dry forest and planted to cotton (Smith and Reynolds 1972). These systems can be extremely productive but often have significant pest problems (see Sect. 9.4).

Traditional annual polycultures typically include two to ten species growing in complex arrangements that incorporate both spatial and temporal heterogeneity (Vandermeer 1989). A large proportion of food crops is produced in such multiple cropping systems in the tropics (Francis 1986). Moreover, the genetic diversity of traditional crops is often strikingly high compared with those in commercial agricultural systems in both the tropics and the temperate zone.

Perennial crops, such as banana or African oil palm, may also be grown in monocultures, or they may be grown with a diverse shade overstory, as was traditionally done with coffee and cacao. Traditionally, coffee and cacao were grown under cover of various species of large shade trees, and the understory contained a diverse assemblage of herbaceous plants, including annual crops (Perfecto and Snelling 1995). Humidity, light penetration, and the leaf litter layer were similar to natural forest. In some parts of the humid tropics, monocultures of newer coffee varieties produced under full sun, which are strikingly different in structure due to the lack of dense canopy cover, are replacing traditional coffee systems (Perfecto and Snelling 1995).

9.2.5 Traditional Rice Systems

In contrast to many traditional agricultural systems, paddy rice systems have extremely low plant species diversity, and high primary and secondary productivity. Yet in two respects paddy fields are biologically diverse. First, like many traditional systems, hundreds of varieties are planted by small farmers. Genetic diversity of the rice probably reduces losses to pests and pathogens (Roger et al. 1991).

Second, traditional rice paddies contain a striking number of aquatic species. In a year-long study in northeast Thailand, Heckman (1979) found some 590 species in a single traditional rice field including 166 species of algae, 83 species of ciliates, 146 arthropod species, 18 fish species, and 10 species of reptiles and amphibians. Thus, although paddies are monocultures, the diversity of aquatic organisms is quite high. Roger et al. (1991)

suggested that biodiversity has apparently decreased in rice fields since the mid-1970s, probably because of increased crop intensification, and the widespread application of pesticides.

9.3 Plant Diversity and Primary Productivity

9.3.1 Comparisons Between Natural and Managed Ecosystems

Managed ecosystems, despite their lower species diversity, often have rates of primary productivity that are equal to or higher than those of adjacent natural ecosystems (Murphy 1975: Loucks 1977). Because ecosystems are often managed for high productivity, water and nutrient subsidies result in production. Comparisons between managed and unmanaged ecosystems of net primary productivity need to account for these differences in subsidies. Managed ecosystems are more appropriately compared with natural ecosystems that receive water or nutrient subsidies, such as marshes and floodplains, rather than natural ecosystems that do not receive high subsidies (Mitchell 1984). Tropical rice ecosystems and tropical marshes, both of which depend heavily on subsidies, differ little in productivity (Mitchell 1984).

High productivity of managed ecosystems sometimes occurs without significant subsidies of nutrients or water. In Puerto Rico, pine and mahogany plantations managed without water or nutrient inputs have lower species richness but greater annual aboveground biomass production than secondary forests of similar ages (Lugo 1992). There was no correlation between species richness and leaf biomass (standing crop) in nine tropical ecosystems studied in Costa Rica and Mexico (Ewel et al. 1982). A largely monocultural plantation of *Gmelina arborata*, a fast-growing timber species, had highest leaf biomass. Moreover, shaded coffee, a mixed planting of cacao-plantain-*Cordia*, and a wooded home garden system all had greater leaf biomass than the species rich secondary forest. However, because the proportion of total biomass allocated to leaves versus woody tissues probably differs among these ecosystems, extrapolation from leaf biomass to total biomass or total productivity is not warranted.

Primary productivity in a four-species mixture of annual and perennial crops in Venezuela was higher during the first year but lower during the second year than productivity of diverse secondary succession vegetation (Uhl and Murphy 1981). The low productivity of the successional vegetation during the first year was attributed to the negative impacts of burning on seed banks and seedling establishment. Because this is one of the few studies that have attempted to compare productivity of natural ecosystems and agroecosystems in the tropics, no general patterns are yet evident with respect to the link between diversity and productivity.

Patterns of leaf area index (LAI) and light interception in different natural and agricultural systems are not necessarily correlated with species richness or diversity. In experimental systems in Costa Rica, Brown and Ewel (1987) found higher LAI values in natural secondary succession than in monocultures of annual agricultural crops. However, LAIs did not differ between secondary succession and a monoculture of the tree *Cordia alliodora* after 4 years (Brown and Ewel 1987). In nine tropical ecosystems including monocultures, agroforestry systems, and secondary forests, Ewel et al. (1982) found LAI values ranging from 1.0 to 5.1, but these values were not related to species diversity. For example, secondary forest and gmelina plantations had equally high LAIs of 5.1. Light penetration to the soil was similar to that in secondary forest in gmelina plantations and sweet potato monocultures, indicating effective light capture in these low diversity systems. These data suggest that the efficiency of light capture within an ecosystem is not strongly correlated with species richness.

Nutrient capture, on the other hand, may be related to species diversity. Root biomass, especially the biomass of fine roots near the soil surface, does seem to increase with species richness. In Puerto Rico, root densities and root biomass were greater in secondary forests than in plantations (Lugo 1992). Ewel et al. (1982) reported higher total root biomass and higher root surface area in systems with higher species richness, such as secondary forest, shaded coffee, a cacao-plantain-*Cordia* system, and a home garden. This relationship between root biomass and species richness appears to result in higher soil nutrient retention under more diverse plant communities (Ewel et al. 1991). However, where monocultures produce extensive root systems, such as the fine root system of *Cordia* (Berish and Ewel 1988), soil nutrient retention may not necessarily be strikingly lower than under diverse secondary successional forests (Ewel et al. 1991).

In traditional coffee plantations with a diverse overstory of shade trees, peak nitrogen demand by coffee appears to be synchronized with nitrogen transfer from shade tree litter to soil (Aranguren et al. 1982). Nitrogen released from rapid decomposition of this litter is likely to be taken up by the coffee whose roots are concentrated in the upper 30 cm of soil and in the litter layer itself. Moreover, traditional shade trees include a number of nitrogen-fixing legumes such as *Inga* spp. and *Erythrina* spp. Compared with coffee litter, shade tree litter was much more abundant, had higher average nitrogen content, and decomposed more slowly. This suggests that adding shade trees to a coffee monoculture is likely to have significant impacts on nitrogen cycling and retention. Shade trees also appear to play an important role in nutrient cycling in cacao and tea plantations (Willey 1975).

In the Venezuelan Amazon, net nitrogen mineralization and nitrification rates, as well as total soil N, were lower in undisturbed forest than in a swidden field dominated by cassava or in swidden pasture dominated by the introduced grass *Brachiaria decumbens* (Montagnini and Buschbacher 1989). If this increase in available N is not fully taken up by crops, the potential for increased leaching and denitrification from these swidden sites is high. The diverse forest system appears to have higher nitrogen retention capacity than the agricultural systems, due at least in part to the thick root mat near the soil surface that can take up nutrients before they leach into the subsoil.

9.3.2 Productivity of Diverse Cropping Systems

The majority of basic grains and root crops are produced in multiple cropping systems that are common in the tropics. In Nigeria, for example, 76% of maize, 80% of cotton, 86% of cocoyam, 90% of millet, 95% of peanut, and 99% of cowpea are produced in polycultures (Okigbo and Greenland 1976). Agriculture in tropical forest zones in Nigeria typically uses intercrops, including various combinations of yams, cassava, maize, cocoyams, bananas, and plantains, whereas in the semiarid regions, millet, sorghum, maize and grain legumes are commonly intercropped. In Latin America as a whole, 60% of maize is grown in polyculture, as is 80% of beans in Brazil and 90% of beans in Colombia (Francis 1986). In the Asian tropics, the widespread cultivation of paddy rice has led to a decline in the polycultures of root, vegetable, and tree crops that were common before the domestication of rice (Plucknett and Smith 1986). In all tropical regions, it is common to intercrop tree crops such as coffee, cacao, rubber, coconut, and oilpalm with annual food crops.

Many investigators have examined the characteristics of species-rich cropping systems and their potential yield advantages over species-poor cropping systems during the past two decades (Trenbath 1974, 1976; Papendick, et al. 1976; Kass 1978; Willey 1979a, b; Beets 1982; Francis 1986, 1989; Vandermeer 1989). The majority of these studies have compared the yields of mixtures of two crop species with pure stands of single crop species, and most of them have addressed two questions: Do crop mixtures produce higher yields than monocultures? Do crop mixtures have greater stability of yield than monocultures? Because crops often are grown for a particular plant part, such as seed, fruit, or tuber, "yield" is seldom synonymous with primary production. Nevertheless, because numerous investigators have measured total aboveground plant biomass, these data can be used to evaluate the influence of crop diversity on aboveground primary productivity. Moreover, for species in which the harvest index (proportion of production allocated to yield) is well characterized, yield can be converted to total productivity (Mitchell 1984). Little information is available on belowground production in most systems, with the result that predictions about the influence of diversity on productivity are limited to aboveground productivity. Unfortunately, there is no simple, reliable way to estimate net primary production from aboveground net productivity (Mitchell 1984).

Crop mixtures, both mixtures of species and mixtures of genotypes within a species, usually have higher productivity than the average of their component species grown in monocultures. In a review of 344 controlled experiments on biomass production of herbaceous mixtures compared to monocultures, Trenbath (1974) reported that 60% of mixtures had higher production than the average of their component crops grown in pure stands, i.e., they exhibited "overyielding." Furthermore, 24% of mixtures had greater production than the monoculture with the highest production, whereas only 13% had lower production than the monoculture with the lowest production. This analysis excluded mixtures of leguminous and

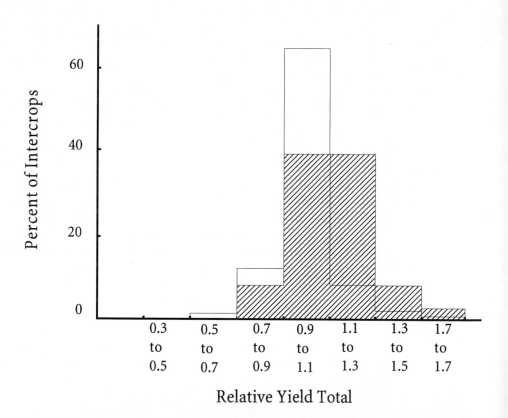

Fig. 9.1. Distributions of relative yield totals for intercrops including (*hatched bars*) or not including (*open bars*) legume-nonlegume combinations. (After Trenbath 1976). The relative yield total measures the yield of an intercrop relative to the yields of the component crops. A relative yield total greater than 1.0 indicates that the intercrop is overyielding in comparison to the monocultures. A relative yield total less than 1.0 indicates that the intercrop is underyielding in comparison to the monocultures

nonleguminous plants; mixtures including both legumes and nonlegumes were even more likely to overyield relative to the component monocultures (Fig. 9.1; Trenbath 1976). These data suggest that an increase in plant species richness is more likely to lead to an increase in primary production, than a decrease, but that an increase in productivity is not inevitable. Clearly, much depends on the particular species combinations and the interactions between the species.

Numerous experimental studies have examined these mechanisms that result in overyielding in species mixtures, such as more efficient use of resources (light, water, nutrients) or reduced pest damage. In general, when interspecific competition is less than intraspecific competition for some limiting factor, then overyielding is predicted (Francis 1989). Also one crop may modify the environment in a way that benefits a second crop by lowering the population of a critical herbivore or releasing nutrients that can be taken up by the second crop (Vandermeer 1989). Facilitation may result in overyielding even where direct competition between crops is substantial.

Light interception may be more efficient in a mixture of crops of different growth forms and temporal patterns of growth. Spatial organization of plants affect total light interception, but increasing the density of a single species can have the same effect as adding a second species. Light use efficiency may be enhanced by the combination of a tall C_4 grass with a short, dense C_3 crop, exemplified by the classic maize/bean polyculture that is widespread in tropical America (Willey 1979a, Francis 1986). The strongest effect of mixtures on light interception probably results when leaf area duration is increased if the growing seasons of the mixed crops only partly overlap. These data indicate that the efficiency of light capture within an ecosystem is not strongly correlated with species richness per se, but may be related to particular characteristics of species combinations.

Nutrient relations in intercrops are complex. Intercrops apparently are more likely to overyield at low nutrient levels (Francis 1989); that is, under conditions of nutrient stress, crop diversity significantly increases productivity. In addition, numerous studies have shown greater nutrient capture in polycultures than in monocultures (reviewed in Willey 1979a; Francis 1989). In the legume-nonlegume intercrops that are traditional throughout the tropics, nitrogen excreted by the legume or made available through decomposing roots may be taken up by the nonlegumes (Willey 1979a). The highest levels of overyielding are often seen in these mixtures, where adding a nitrogen-fixing legume can increase the overall yield of the non-legume crop. In Costa Rica, for example, Rosset et al. (1984) reported higher tomato yields in tomato-bean intercrops than in tomato monocultures where tomato density was held constant. Moreover, the addition of beans reversed the negative effects of increasing tomato plant density on tomato yields: yields of high density monocultures were lower than those of low density monocultures, whereas yields of high density polycultures were

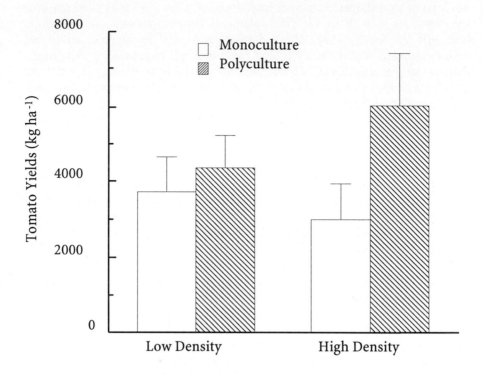

Fig. 9.2. Tomato yields (kg ha^{-1}) in tomato monocultures and tomato-bean polycultures with high and low densities of tomatoes. (Rosset et al. 1984)

higher than those of low density polycultures (Fig. 9.2). In this case, yields represent biomass allocated to reproductive tissue (fruits), rather than total plant biomass.

Although interest in polycultures has concentrated on the impacts of adding crop species to monocultures, the addition of animals may also have significant consequences for ecosystem processes. Asian fish-rice systems have received increasing attention in recent years, because preliminary studies have demonstrated higher rice yields in systems that include fish (Lightfoot et al. 1993). Lightfoot et al. (1993) modeled productivity and nutrient cycling in fish-rice polycultures. Model results indicated that, with the addition of fish to rice monocultures, rice yields and soil microbial biomass would increase, and the biomass of invertebrates and aquatic weeds would decrease. Lightfoot et al. (1993) suggested several mechanisms to explain the positive effects of fish on rice, including faster nutrient turnover and increased nitrogen availability. Preliminary studies also indicate that fish can significantly reduce insect pest populations as well as weed biomass. Results from this model are intriguing, but await empirical tests.

9.3.3 Stability of Diverse Cropping Systems

Conventional wisdom suggests that diverse cropping systems not only tend to have higher yields but also tend to have more stable yields in the face of environmental variability. Rao and Willey (1980) describe three mechanisms that might lead to yield stability. First, when one crop performs poorly, because of drought or pest epidemic for example, the other crop(s) can compensate, using the space and resources made available. Second, if the yield advantages of intercrops are greater under stress conditions, then yield stability is higher. Finally, where intercropping leads to reduced pest attack, as it often does (see Sect. 9.4), then greater yield stability may result. In contrast to Rao and Willey's predictions about yield stability, Vandermeer and Schultz (1990) have argued on theoretical grounds that higher yield stability is expected in intercrops only where facilitation between the crops is operating; where the mechanism for overyielding is reduced competition, then intercrop yields are expected to be less stable than monoculture yields.

In practice, the empirical evidence for yield stability in intercrops is suggestive but equivocal. Natarajan and Willey (1986) examined the effect of drought on polyculture overyielding by manipulating water stress on intercrops of sorghum and peanut, millet and peanut, and sorghum and millet. Although total biomass production in both polycultures and monocultures decreased as water stress increased, all of these intercrops overyielded consistently at five levels of moisture availability ranging from 297 to 584 mm of water applied over the cropping season. Interestingly, the rate of overyielding actually increased with water stress, such that the relative differences in productivity between monocultures and polyculture became more accentuated as stress increased. These data appear consistent with the idea that species richness buffers productivity under conditions of environmental variability (Tilman and Downing 1994).

Rao and Willey (1980) reported higher yield stability of 51 sorghum/pigeonpea intercrops compared with the stability of sorghum and pigeonpea monocultures in India. Yield stability was measured in three ways: by calculating coefficients of variation, by computing regressions of yield against an environmental index, and by estimating the probability of crop failure. Based on all measures, intercrops exhibited greater yield stability: intercrops had lower coefficients of variation than either sorghum or pigeonpea separately; the response of intercrops to environmental change was as stable or more stable than the most stable component crop (sorghum); and intercrops showed a much lower probability of failure than either of the component crops. This latter measure is effectively an estimate of risk and results from both the higher yields of intercrops and the putative yield stability. Although the study by Rao and Willey (1980) is arguably the most comprehensive examination of intercrop yield stability to date, Vandermeer and Schultz (1990) suggest that there were methodological problems with the analysis of these data and that the question of yield stability is still open.

9.4 Plant Diversity and Primary Consumers

The impact of plant diversity on primary consumers can be examined by asking how plant diversity influences: herbivore standing stock and secondary production; and rates of herbivory (i.e., plant tissue loss). The first influence can be assessed by measuring herbivore load, the total biomass of herbivores per unit plant material (Root 1973). Surprisingly, since Root (1973) first reported lower herbivore loads on collards in diverse vegetation compared to collards in monoculture, herbivore load has rarely been measured in any comparison of herbivores in diverse and simple plant communities. Moreover, there appear to be no published studies addressing herbivore load or secondary production in tropical systems.

A more common way of assessing the impact of plant diversity on primary consumers is to compare herbivore densities in diverse and simple plant communities. In general, arthropod herbivores are less abundant but more species-rich in diverse cropping systems than in monocultures (Risch et al. 1983; Andow 1991). Andow's (1991) review of 209 studies of 287 species of herbivorous arthropods, both temperate and tropical, determined that 52% of the species had consistently lower densities in polycultures, whereas only 15% had consistently higher densities in polycultures; the remainder showed no response or a variable response to increased plant diversity (Fig. 9.3A). However, specialist herbivores (i.e., monophagous herbivores that fed on a single crop in the polyculture) and generalist herbivores (polyphagous herbivores that fed on more than one crop in the polyculture) responded differently (Fig. 9.3B). Densities of specialists were more likely to decrease (59% of species studied) in polycultures than to increase (8% of species studied), whereas generalists were more likely to increase (40%) than decrease (28%). Densities of specialists may decrease because: (1) they have difficulty locating hosts in a diverse system, possibly due to interference with olfactory or visual cues; or (2) they leave hosts more often due to lower plant quality and then have difficulty relocating them. Generalists that feed on all components of the polyculture do not encounter these difficulties. Thus herbivore host range is an important factor determining arthropod response to plant diversity (Andow 1991). However, lower herbivore densities in diverse systems appear to be a general result.

The second approach to examining the impact of plant diversity on primary consumers compares herbivory rates in plant communities of varying diversity. Ewel et al. (1982) measured leaf damage in nine tropical ecosystems and found no strong relationship between species diversity and ecosystem-level measurements of damage (e.g., whole-ecosystem leaf area loss and whole-ecosystem biomass loss). Leaf area loss was lowest in the sweet potato monoculture, which received insecticide treatments, and highest in the mature maize. Leaf damage to individual species was some

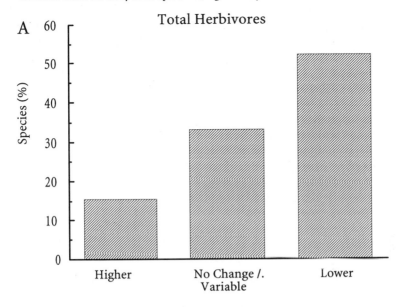

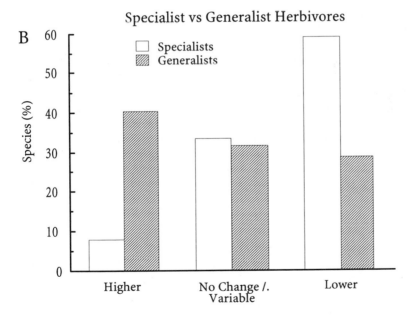

Fig. 9.3A,B. Percent of herbivorous arthropod species that have either higher, lower, or variable, or unchanged population densities in polycultures compared to monocultures. For convenience, variable and unchanged responses are grouped together in a single category. Data are based on 287 species (Andow 1991). **A** All herbivorous arthropods. **B** Specialist (monophagous) herbivorous arthropods versus generalist (polyphagous) herbivorous arthropods

times, but not always, lower in diverse ecosystems. However, these data are difficult to interpret, since the time interval over which herbivory occurred varied among the ecosystems. In a more controlled study of herbivory rates in different ecosystems, Brown and Ewel (1987) found a trend for lower losses per unit leaf area in the species-rich ecosystems, although total leaf biomass losses to herbivory did not differ significantly among ecosystems. Furthermore, the temporal variability of leaf loss was significantly lower in the diverse ecosystems than in the monocultures.

In analyzing the influence of plant diversity on herbivory, yield loss can be used as a surrogate for herbivory rates (Andow 1991). We are aware of only one tropical study that collected data suitable for such an analysis. In Colombia, common bean (*Phaseolus vulgaris*) yields in monocultures and polycultures with and without the leafhopper *Empoasca kraemeri* were compared by applying insecticides to some plots (Schoonhoven et al. 1981). In the treatments without insecticides, leafhopper densities were consistently lower in polycultures than in monocultures. In four trials using a bean variety that was susceptible to leafhopper damage, both absolute and proportional yield loss was less in the polycultures than in the mono-culture, consistent with a lower herbivory rate in polyculture. In contrast, four trials using a bean variety that was tolerant to leafhopper feeding indicated that absolute yield loss was similar or lower in polycultures, but proportional yield loss was similar or slightly higher in polycultures compared to monocultures. Using yield loss as a surrogate for herbivory rate is thus complicated by variability in plant tolerance to insect damage. Nevertheless, the overall results suggest that herbivory is lower in polycultures, especially when plants display little tolerance to herbivore damage.

In summary, the evidence suggests that plant species richness has important consequences for both herbivore standing stock and rates of herbivore attack. The many studies on insect population densities in monocultures and polycultures support the notion that herbivory is lower in systems with higher plant species richness. Very few studies directly address herbivore standing stock, or secondary production as a function of plant species richness, but the available evidence suggests that all these may decrease as plant diversity increases. Although studies on rates of herbivory are more equivocal, the loss of plant reproductive tissues by herbivores and the temporal variability of attack appear to be inversely related to plant diversity.

9.5 Plant Diversity and Secondary Consumers

In his comprehensive review of investigations of the influence of plant diversity on both temperate and tropical insect populations, Andow (1991) reported that 53% of natural enemy species were more abundant in poly

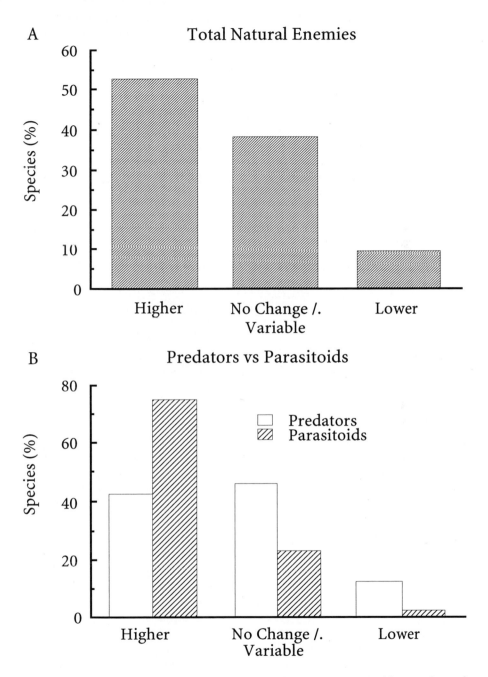

Fig. 9.4A,B. Percent of natural enemy species that have higher, lower, variable or unchanged population densities in polycultures compared to monocultures. For convenience, variable and unchanged responses are grouped together into a single category. Data are based on 130 species (Andow 1991). **A** All natural enemies. **B** Predators versus parasitoids

cultures than in monocultures, whereas only 9% of the species were less abundant (Fig. 9.4A). The remainder showed no response or a variable response to plant diversity. Parasitoids were particularly likely to increase (75% of species studied) rather than decrease (3% of species studied) in polycultures. Predators were more variable than parasitoids, although they were still more likely to increase than decrease in polycultures (Fig. 9.4B). Moreover, predation rates, parasitism rates, and the ratio of natural enemies to herbivores all tend to be higher in polycultures than in monocultures (Russell 1989). These data indicate that natural enemies typically maintain higher populations in diverse systems, but the mechanisms underlying this response to plant diversity are not yet well understood (Russell 1989; Andow 1991).

9.5.1 Ants in Diverse Cropping Systems

Ants are among the most important generalist predators in tropical habitats, including both agricultural and forest ecosystems. Ants are both numerically and ecologically dominant in tropical forests and perennial tree plantations, where the non-overlapping territories of dominant ants form a three-dimensional mosaic that significantly influences the distribution and abundance of other arthropods (Majer 1993). The "ant mosaic" is a key biotic organizational feature that influences the diversity and species composition of neotropical forest biota, including plants and vertebrates as well as invertebrates (Gilbert 1980). In Ghanaian cacao plantations ants can constitute 89% of the total insect numbers (Leston 1973) and up to 70% of arthropod biomass (Majer 1976). They comprise 10-33% of the arboreal arthropod biomass in Brazilian cacao plantations (Majer et al. 1994). In addition to the dominance of ants in tropical perennial crops, recent studies support the notion that the ant assemblage also plays a significant role in controlling herbivorous insects in tropical annual agroecosystems as well (Risch and Carroll 1982b; Perfecto 1991; Perfecto and Sediles 1992). Risch and Carroll (1982b) found significant increases in the diversity and abundance of both herbivores and non-ant predators when *Solenopsis geminata* was excluded from annual agroecosystems.

Despite the importance of ants as predators in tropical ecosystems, there is little information about the influence of plant diversity on ant foraging activity. Nestel and Dickschen (Nestel and Dickschen 1990) studied the foraging efficiency of ground-foraging ants in coffee agroecosystems with and without shade trees in Mexico. Ants discovered tuna baits more quickly in coffee systems with no shade trees compared to those with either a single species of shade tree or multiple species. In the unshaded coffee systems, the ant fauna was dominated by a single abundant, highly aggressive ant species, *S. geminata*, whereas other ant species were also present in the shaded system.

Studies of the influence of plant diversity on ant abundance or predation in annual systems have reported mixed results. Perfecto and Sediles (1992) report no significant difference in ant foraging activity between maize monocultures, bean monocultures, and maize-bean polycultures in Nicaragua. In Mexico, Letourneau (1987) found significantly higher ant activity in squash monoculture compared to squash-maize-cowpea polycultures. However, other studies in Mexico have suggested that crop cultivation history interacts with cropping system to determine ant foraging activity (Saks and Carroll 1980). In comparisons of maize-bean-squash polycultures and maize and bean monocultures, Saks and Carroll (1980) found highest rates of ant foraging activity in polycultures that had been planted continuously for several years, whereas foraging activity was lower in monocultures and in newly planted polycultures.

The diversity of vegetation surrounding the agroecosystem system may also be an important determinant of ant activity and predation (Risch and Carroll 1982a). Ground-foraging ants were sampled in two corn/bean /squash polycultures in Mexico, one surrounded by 40-year-old forest (the forest *milpa*) and one surrounded by 1-year-old secondary growth (the field *milpa*). Ant foraging efficiency, measured by the occupancy and removal rates of baits, was higher in the field *milpa* than the forest *milpa* (Risch and Carroll 1982a). The forest *milpa* had a relatively diverse ant assemblage but the ant assemblage of the field *milpa* was dominated by *Solenopsis geminata*, which prefers open habitats and colonizes new agricultural fields rapidly. The forest *milpa* supported higher biodiversity than the field *milpa*, but the predation rate was higher in the agroecosystem surrounded by fields rather than forest. Given that the 40-year-old forest was presumed to have greater plant species richness than secondary growth, and was determined to have higher ant species richness, these data suggest that plant diversity does not necessarily lead to higher predation rates by ants.

9.6 Conclusions

The influence of biodiversity on the functioning of tropical managed ecosystems is undoubtedly complex, and conclusions about the relationship between diversity and ecosystem functioning differ according to the variables being measured and the types of managed ecosystems under study. There is no simple correlation between plant species richness and primary productivity. Rates of primary production are not necessarily lower in managed, species-poor ecosystems than in natural, species-rich ecosystems. In addition, the efficiency of light capture and leaf area index are not always correlated with species richness (Willey 1979a; Ewel et al. 1982; Brown and Ewel 1987), but there is some evidence that the efficiency of nutrient capture and retention may increase with species richness (Willey 1979a; Ewel et al. 1982; Lugo 1992).

In managed ecosystems, such as annual cropping systems with low numbers of plant species, plant species richness can have significant impacts on primary production. Polycultures generally overyield with respect to monocultures (Trenbath 1974; Willey 1979a,b), and there is evidence that yields are more stable in the face of environmental variability (Rao and Willey 1980). Higher yields in polycultures may result from more efficient use of resources (e.g., light, water, nutrients) in time or space, reduced pest attack, or facilitation between crops. These yield advantages in more diverse systems appear to be particularly pronounced under conditions of stress, such as drought or nutrient deficiencies. Furthermore, arthropod herbivores, and in particular specialist consumers, are generally less abundant in more diverse cropping systems, whereas natural enemies of insect pests are generally more abundant. Studies documenting how plant diversity influences rates of herbivory in tropical managed ecosystems are more equivocal, although the loss of plant reproductive structures and variability of temporal attack appear to be lower in more diverse cropping systems.

Many of the studies examined here were conducted in cropping systems with low plant diversity. At present, we know little about functional redundancy in such systems. For example, it is unclear whether a polyculture with ten species would display proportionately greater overyielding or reduced pest attack compared to a polyculture with three crop species. Can lessons from annual cropping systems be applied to home gardens and managed forests? Because plant species diversity is manipulated in managed ecosystems, they provide great potential for testing hypotheses about biodiversity and ecosystem functioning. However, many existing studies that we review here have been designed to maximize information on yields of particular plant products. In the future, we need to design studies with the explicit goal of understanding the link between biodiversity and ecosystem processes of tropical managed ecosystems.

References

Alcorn J (1990) Indigenous agroforestry systems in the Latin American tropics. In: Altieri MI, Hecht SB (eds) Agroecology and small farm development. CRC Press, Baton Rouge, pp 203-211

Alcorn JB (1983) El te'lom huasteco: presente, pasado, y futuro de un sistema de silvicultura indígena. Biotica 8:315-325

Allison J (1983) An ecological analysis of home gardens (*huertos familiares*) in two Mexican villages. Univ California, Santa Cruz

Andow DA (1991) Vegetational diversity and arthropod population response. Annu Rev Entomol 36:561-586

Aranguren J, Escalante G, Herrera R (1982) Nitrogen cycle of tropical perennial crops under shade trees. Plant Soil 67:247-258

Barrera A, Gómez-Pompa A, Yanes CV (1977) El manejo de las selvas por los Mayas: Sus implicaciones silvícolas y agrícolas. Biotica 2:47-61

Beets WC (1982) Multiple cropping and tropical farming systems. Westview Press, Boulder

Berish CW, Ewel JJ (1988) Root development in simple and complex tropical successional ecosystems. Plant Soil 106:73-84

Boster J (1983) A comparison of the diversity of Jivaroan gardens with that of the tropical forest. Hum Ecol 11:47-68

Brown BJ, Ewel JJ (1987) Herbivory in complex and simple tropical successional ecosystems. Ecology 68:108-116

Brownrigg L A (1985) Home gardening in international development. League for Int Food Education, US Agency Int Dev, Washington DC

Christanty L, Abdoellah OE, Marten GG, Iskander J (1986) Traditional agroforestry in West Java: the *pekarangan* (homegarden) and *kebun-talun* (annual-perennial rotation) cropping systems. In: Marten GG (ed) Traditional agriculture in Southeast Asia. A human ecology perspective. Westview Press, Boulder, pp 132-138

Denevan WM, Tracy JM, Alcorn JB, Padoch C, Denslow JS (1984) Indigenous agroforestry in the Peruvian Amazon: Bora Indian management of swidden fallows. Interciencia 9:346-357

Ewel JJ, Gliessman S, Amador M, Benedict F, Berish C, Bermudez R, Brown B, Martinez A, Miranda R, Price N (1982) Leaf area, light transmission, roots, and leaf damage in nine tropical plant communities. Agro-Ecosystems 7:305-326

Ewel JJ, Mazzarino MJ, Berish CW (1991) Tropical soil fertility changes under monocultures and successional communities of different structure. Ecol Appl 1:289-302

Francis CA (1986) Introduction: distribution and importance of multiple cropping. In: Francis, CA (ed) Multiple cropping systems. Macmillan, New York, pp 1-19

Francis CA (1989) Biological efficiencies in multiple cropping systems. Adv Agron 42:1-36

Gilbert LE (1980) Food web organization and conservation of neotropical diversity. In: Soulé ME, Wilcox BA (eds) Conservation biology. An evolutionary-ecological perspective. Sinauer, Sunderland, Massachusetts, pp 11-33

Gliessman SR (1990a) Integrating trees into agriculture: the home garden agroecosystem as an example of agroforestry in the tropics. In: Gliessman SR (ed) Agroecology. Researching the ecological basis for sustainable agriculture. Springer, Berlin Heidelberg New York, pp 160-168

Gliessman SR (1990b) Understanding the basis of sustainability for agriculture in the tropics: experiences in Latin America. In: Edwards CA, Lal R, Madden P, Miller RH, House G (eds) Sustainable agricultural systems. Soil Water Conserv Soc, Ankeny, Iowa, pp 170-190

Gómez-Pompa A (1987) On Maya silviculture. Mex Stud 3:1-17

Gómez-Pompa A, Flores JS, Sosa V (1987) The "Pet Kot": a man-made tropical forest of the Maya. Interciencia 12:10-15

Gordon BL (1982) A Panama forest and shore. Natural history and Amerindian culture in Bocas del Toro. Boxwood Press, Pacific Grove, California

Heckman CW (1979) Rice field ecology in northeastern Thailand. Dr W Junk, Boston

Herrera-Castro N (1991) Los huertos familiares Mayas en el oriente de Yucatan. Univ Nac Autónoma de México, México, DF

Kass DC (1978) Polyculture cropping systems: review and analysis. Cornell Int Agric Bull 32:1-6.

Kimber C (1966) Dooryard gardens of Martinique. Yearb Assoc Pac Coast Geogr 28:97-118

Leston D (1973) The ant mosaic - tropical tree crops and the limiting of pests and diseases. Pest Abstr News Summaries 19:311-341

Letourneau DK (1987) The enemies hypothesis: tri-trophic interactions and vegetational diversity in tropical agroecosystems. Ecology 68:1616-1622

Lightfoot C, Roger PA, Cagauan AG, de la Cruz CR (1993) Preliminary steady-state nitrogen models of a wetland ricefield ecosystem with and without fish. In: Christensen V, Pauly D (eds) ICLARM Conf Proc 26, Manila, pp 56-64

Loucks OL (1977) Emergence of research on agro-ecosystems. Annu Rev Ecol Syst 8:173-192

Lugo A (1992) Comparison of tropical tree plantations with secondary forests of similar age. Ecol Monogr 62:1-41

Majer JD (1976) The ant mosaic in Ghana cocoa farms: further structural considerations. J Appl Ecol 13:145-156.

Majer JD (1993) Comparison of the arboreal ant mosaic in Ghana, Brazil, Papua New Guinea and Australia: its structure and influence on arthropod diversity. In: Lasalle J, Gauld ID (eds) Hymenoptera and biodiversity. CAB International, Wallingford, pp 115-141

Majer JD, Delabie JHC, Smith MRB (1994) Arboreal ant community patterns in Brazilian cocoa farms. Biotropica 26:73-83

May RM (1984) Prehistory of Amazonian Indians. Nature 312:19-20

Mitchell R (1984) The ecological basis for comparative primary production. In: Lowrance R, Stinner BR, House GJ (eds) Agricultural ecosystems. Unifying concepts. Wiley, New York, pp 13-53

Montagnini F, Buschbacher R (1989) Nitrification rates in two undisturbed tropical rain forests and three slash-and-burn sites of the Venezuelan Amazon. Biotropica 21:9-14

Murphy PG (1975) Net primary productivity in tropical terrestrial ecosystems. In: Lieth H, Wittaker RH (eds) Primary productivity of the biosphere. Springer, Berlin Heidelberg New York, pp 217-231

Natarajan M, Willey RW (1986) The effects of water stress on yield advantages of inter-cropping systems. Field Crops Res 13:117-131

National Research Council (1993) Sustainable agriculture and the environment in the humid tropics. Natl Acad Press, Washington DC

Nestel D, Dickschen F (1990) The foraging kinetics of ground ant communities in different mexican coffee agroecosystems. Oecologia 84:58-63

Okigbo BN, Greenland DJ (1976) Intercropping systems in tropical Africa. In: Papendick RI, Sanchez PA, Triplett GB (eds) Multiple cropping. Am Soc Agron, Madison, Wisconsin, pp 63-101

Papendick RI, Sanchez PA, Triplett GB (1976) Multiple cropping. Am Soc Agron, Madison, Wisconsin

Perfecto I (1991) Ants (Hymenoptera: Formicidae) as natural control agents of pests in irrigated maize in Nicaragua. J Econ Entomol 84:65-70

Perfecto I, Sediles A (1992) Vegetational diversity, ants (Hymenoptera: Formicidae), and herbivorous pests in a neotropical agroecosystem. Environ Entomol 21:61-67

Perfecto I, Snelling R (1995) Biodiversity and tropical ecosystem transformation: ant diversity in the coffee agroecosystem in Costa Rica. Ecological Applications, (in press)

Pimentel D, Stachow U, Takacs DA, Brubaker HW, Dumas AR, Meaney JJ, O'Neil JAS, Onsi DE, Corzilius DB (1992) Conserving biological diversity in agricultural/forestry systems. BioScience 42:354-362

Pinkley HV (1973) The Ethnoecology of the Kofan Indians. Harvard Univ, Cambridge

Plucknett DL, Smith NJH (1986) Historical perspectives on multiple cropping. In: Francis CA (ed) Multiple cropping systems. Macmillan, New York, pp 20-39

Posey DA (1985) Indigenous management of tropical forest ecosystems: the case of the Kayapo Indians of the Brazilian Amazon. Agrofor Syst 3:13.

Rao MR,Willey RW (1980) Evaluation of yield stability in intercropping: studies on sorphum/pigeonpea. Exp Agric 16:105-116

Reid WV, Miller KR (1989) Keeping options alive: the scientific basis for conserving biodiversity. World Resour Inst, Washington DC

Rico-Gray V, Gómez-Pompa A, Chan C (1985) Las selvas manejadas por los mayas de Yohaltun, Campeche, México. Biotica 10:321-327

Risch SJ, Carroll CR (1982a) The ecological role of ants in two Mexican agroecosystems. Oecologia 55:114-119

Risch SJ, Carroll CR (1982b) Effect of a keystone predaceous ant, *Solenopsis geminata*, on arthropods in a tropical agroecosystem. Ecology 63:1979-1983

Risch SJ, Andow D, Altieri MA (1983) Agroecosystem diversity and pest control: data, tentative conclusions, and new research directions. Environ Entomol 12:625-629

Roger PA, Heong KL, Teng PS (1991) Biodiversity and sustainability of wetland rice production: role and potential of microorganisms and invertebrates. In: Hawksworth DL (ed) The biodiversity of microorganisms and invertebrates: its role in sustainable agriculture. CAB International, Wallingford, pp 117-136

Root RB (1973) Organization of a plant-arthropod association in simple and diverse habitats: the fauna of collards (*Brassica oleracea*). Ecol Monogr 43:95-194

Rosset PM, Ambrose RJ, Power AG, Hruska AJ (1984) Overyielding in polycultures of tomato and bean in Costa Rica. Trop Agric (Trinidad) 61:208-212

Russell EP (1989) Enemies hypothesis: a review of the effect of vegetational diversity on predatory insects and parasitoids. Environ Entomol 18:590-599

Saks ME, Carroll CR (1980) Ant foraging activity in tropical agroecosystems. Agro-Ecosyst 6:177-188

Schoonhoven AV, Cardona C, Garcia J, Garzon F (1981) Effect of weed covers on *Empoasca kraemeri* Ross and Moore populations and dry bean yields. Environ Entomol 10:901-907

Smith RF, Reynolds HT (1972) Effects of manipulation of cotton agro-ecosystems on insect pest populations. In: Farvar MT, Milton JP (eds) The careless technology. Ecology and international development. Natural History Press, Garden City, New York, pp 373-406

Tilman D, Downing JA (1994) Biodiversity and stability in grasslands. Nature (London) 367:363-365

Torquebiau E (1984) Man made dipterocarp forest in Sumatra. Agrofor Syst 2:103-127

Trenbath BR (1974) Biomass productivity of mixtures. Adv Agron 26:177-210

Trenbath BR (1976) Plant interactions in mixed crop communities. In: Papendick RI, Sanchez PA, Triplett GB (eds) Multiple cropping. Am Soc Agron, Madison, Wisconsin, pp 129-170

Uhl C, Murphy P (1981) A comparison of productivities and energy values between slash and burn agriculture and secondary succession in the upper Rio Negro region of the Amazon basin. Agro-Ecosyst 7:63-83

Vandermeer J (1989) The ecology of intercropping. Cambridge Univ Press, Cambridge

Vandermeer J, Schultz B (1990)Variability, stability, and risk in intercropping: some theoretical explorations. In: Gliessman SR (ed) Agroecology. Researching the sustainable basis for sustainable agriculture. Springer, Berlin Heidelberg New York, pp 205-229

Western D, Pearl MC (1989) Conservation for the twenty-first century. Oxford Univ Press, New York

Wiersum FF (1982) Tree gardening and taungya on Java: examples of agroforestry techniques in the humid tropics. Agrofor Syst 1:53-70

Wilken GC (1977) Integrating forest and small-scale farm systems in Middle America. Agro-Ecosyst 3:291

Willey RW (1975) The use of shade in coffee, cocoa, and tea. Horticult Abstr 45:791-798

Willey RW (1979a) Intercropping - its importance and its research needs. Part I. Competition and yield advantages. Field Crop Abstr 32:1-10

Willey RW (1979b) Intercropping - its importance and its research needs. Part II. Agronomic relationships. Field Crop Abstr 32:73-85

10 Synthesis

Gordon H. Orians[1], Rodolfo Dirzo[2], and J. Hall Cushman[3]

10.1 Introduction

The preceding chapters discuss some important patterns in relationships between tropical forest biodiversity and ecosystem processes. They also show the great gaps in current knowledge of the nature and strength of most of those relationships. Here we synthesize the cross-cutting patterns and issues that arose several times in those chapters. In addition, we consider a number of research themes, which, if pursued, would improve our understanding of the role of biodiversity in the functioning of tropical forests.

Tropical forest vegetation, which develops under a variety of climates, is rich and varied. Although temperatures are relatively constant throughout the year in tropical regions, mean annual temperatures drop predictably with increasing elevation. In addition, seasonality of rainfall, which characterizes all tropical regions to some degree, exerts strong influences on almost all biological processes and patterns in the tropics. Therefore, to set the stage for our synthesis we first review gradients in moisture, soil fertility, and elevation, the major environmental factors that determine the nature and distribution of different types of tropical vegetation (Walter 1973).

[1] Department of Zoology, University of Washington, Box 351800, Seattle, Washington 98195, USA
[2] Centro de Ecología, Universidad Nacional Autónoma de México, Apartado Postal 70-275, México 04510 D. F
[3] Department of Biology, Sonoma State University, Rohnert Park, California 94928, USA

Ecological Studies, Vol. 122
Orians, Dirzo and Cushman (eds) Biodiversity and Ecosystem Processes in Tropical Forests
© Springer-Verlag Berlin Heidelberg 1996

10.2 Environmental Gradients

10.2.1 Moisture

Associated with a moisture gradient from wet to dry is a corresponding gradient of tropical forests from evergreen to seasonally deciduous. Seasonal drought exerts much stronger influences on herbs and understory shrubs than on canopy trees. Dry season drought increases mortality and gap dependence in *Pleiostachya pruinosa*, a herb that dies soon after canopy closure, on Barro Colorado Island, Panama. This is shown by the fact that individuals survive as well in irrigated forest understory as in control gaps (Mulkey et al. 1991b). Small differences in dry season rainfall can greatly affect photosynthetic potential of understory shrubs (Mulkey et al. 1991a; Wright 1991). The most severe dry season in 70 years of record-keeping on BCI occurred during the 1982-83 El Niño. By 1990 populations of 8 of 25 shrub species (32%) declined more than 30% whereas only 4 of 132 tree species (3%) had comparably high mortality rates (Wright and van Schalk 1994).

Production of leaves, flowers, fruit, and litter is more clumped temporally in regions with long dry seasons than in regions with relatively constant rainfall (Opler et al. 1976; Foster 1982a,b; Lieberman 1982; Leighton and Leighton 1983; Wright 1992). Therefore, the flow of energy occurs in marked pulses in tropical dry forests, and litterfall is concentrated during the dry season, followed by rapid decomposition at the beginning of the wet season (Silver et al., Chap 4, this Vol.). The relative importance of plants that flower and fruit during the dry season probably is also positively correlated with increasing dryness

Tree falls occur more often when tropical soils become saturated after a dry period (Brokaw 1982; Brandani et al. 1988). Drying/wetting cycles also accelerate replenishment of the available soil nitrogen from microbial, recalcitrant, or physically protected nitrogen pools. Fluctuations in soil moisture induce oscillations in populations of soil microbes, resulting in pulses of nutrient release. These cycles in soil nutrient availability and moisture may increase uptake of limiting nutrients by plants (Birch 1958, Lodge et al. 1994). Even relatively brief rainless intervals can cause large reductions in soil moisture. At La Selva, Costa Rica, a site without a well-marked dry season, a 40 % reduction in total soil moisture content in the upper 70 cm of soil followed a 1-month period without significant rainfall. Such reductions are sufficient to cause water stress in the forest vegetation (Sanford et al. 1994). With increasing length of the dry season, the drying of the soil becomes more extreme and the pulsing of microbial populations and nutrient release probably become especially marked, but the influence of this pattern on ecosystem productivity and efficiency of natural nutrient cycling is unknown.

The frequency of fires is likely to increase along a moisture gradient from wet to dry in tropical regions. Because fires affect functional groups differently, they may alter ecosystem processes both directly and indirectly. For example, if understory shrubs are more severely harmed by fires than trees, as is likely to be the case, the resources upon which a large number of pollinators and fruit eaters depend in tropical dry forests may be diminished extensively by fire, altering the energy flow interface at the level of primary consumers. Fires also favor animals able to escape by moving out of the area or by burrowing (Braithwaite 1987).

10.2.2 Fertility

Although belief that all tropical soils are red, infertile, and harden irreversibly when they are cleared is widespread, soils of the lowland tropics are as diverse as those of any other region. Infertile red and yellow oxisols and ultisols are common throughout the tropics, but only about seven percent of the tropical soils are highly infertile (Sánchez 1976). Highly fertile soils are common along rivers and in volcanically active areas. A reduction in species richness, vegetation layers, canopy height, and mean leaf size is associated with very low soil fertility in some tropical forests (Brunig 1983). The data from the few tropical forests that have been studied intensively have been summarized by Jordan (1985). Not surprisingly, aboveground productivity is positively correlated with soil fertility. Published values of productivity of leaf litter range from 11.3 to 4.95 Mg ha^{-1} yr^{-1}. Wood production data are not available from the most fertile sites, but medium fertility sites have nearly double the production of the least fertile site, suggesting that the range in wood production across a fertility gradient is comparable to that for leaves.

Trees on fertile soils produce large quantities of nutrient-rich, nonsclerophyllous foliage that decomposes rapidly. In contrast, trees on infertile soils produce smaller quantities of nutrient-poor, scleromophyllous leaves that decompose more slowly. The rate of leaf decomposition varied by nearly a factor of five along the soil fertility gradient in the sites Jordan (1985) compared. Highly productive tropical forests sustain higher levels of herbivory per unit of primary production than unproductive ones, and insect frass may decompose faster than leaves. Therefore, the relative importance of direct flow of energy to detritivores decreases as ecosystem productivity increases (McNaughton et al. 1989).

Standing aboveground biomass in tropical forests is also positively correlated with fertility, but belowground biomass is inversely correlated with fertility. Published values of root-shoot ratios increase from 0.03 to 0.49 along a gradient from the most to the least fertile sites. Up to 25 % of the roots in the forests growing on infertile soils are in a superficial mat; such mats are less well developed in forests on fertile soils (Sanford 1989; Silver and Vogt 1993).

The relative allocation of plant resources to defense (Coley et al. 1985) and allocation to reproduction (Gentry and Emmons 1987) are inversely correlated with soil fertility. This contributes to the slower decomposition rates of litter in forests on infertile soils and presumably results in smaller populations and lower richness of species of animals that depend on pollen, nectar, and fruits.

10.2.3 Elevation

As elevation increases on the slopes of tropical mountains, temperatures and evaporation decrease, wind, rainfall, and waterlogging of soils may increase, and there is a greater incidence of landslides (Leigh 1975; Lawton and Putz 1988). Correlated with these physical changes are reduced rates of productivity and decomposition.

Few data are available for tropical premontane forests, but litter production at the El Verde forest in Puerto Rico is only half as great as in productive lowland forests (Odum 1970), and wood production is similar to that in lowland forests on highly infertile soils. Slower rates of decomposition result in increased nutrient retention in decomposing litter and slow rates of nutrient transfer. Litter accumulation also dramatically alters the soil surface and the composition of the litter fauna.

Aboveground biomass decreases but belowground biomass increases with elevation in tropical forests. The root mass at El Verde is greater than that of lowland forests on rich soils by about a factor of 6, but aboveground biomass is less by only a factor of 2, in part because the leaf biomass of premontane forests is similar to that of lowland forests.

With increasing elevation, canopy height decreases, palms drop out (except on islands), and there is an overall reduction in leaf sizes (Leigh 1975; Tanner and Kapos 1982; Brown et al. 1983; Gentry 1988). The number of species in most taxa also decreases with elevation (see Terborgh 1977 for birds and Janzen 1987 for insects). In addition, there are shifts in the relative representation of life forms. Vines and lianas become less common and epiphytes increase in abundance, species richness, and structural diversity (Brown et al. 1983), resulting in increased nutrient capture, nutrient retention, and nutrient transfers at the atmosphere-terrestrial interface (Silver et al., Chap. 4, this Vol.).

10.3 Biodiversity and Functioning of Tropical Forests

The extent to which human activities are leading to the extinction of species in tropical forests is uncertain, but current estimates of the rates of loss of tropical moist forests are extremely high (Green and Sussman 1990; Whitmore and Sayer 1992; Wilson 1992; FAO 1993; Skole and Tucker 1993). If continued, they are expected to result in extinctions of many species

(Reid 1992), in part because of the small geographical ranges of many tropical species. For example, many species of cloud forest plants in Latin America are endemic to isolated sites smaller than 10 km^2 (Gentry 1992). Among the birds of South American tropical forests, 440 species (25% of the total) have ranges less than 50 000 km^2. In contrast, only eight species (2% of the total) of bird species in the United States and Canada have such restricted ranges (Terborgh and Winter 1980). However, researchers do not agree on the extent of probable losses and by how much they might be reduced by management practices in tropical forests (Lugo 1988; Lugo et al. 1993). Whatever the degree of loss of species in tropical forests, reductions in species richness can be expected to influence functional properties of tropical forest ecosystems in many ways.

In our analysis of tropical forests we used two major ecosystem processes – energy flow and nutrient cycling – as the primary bases for recognizing functional groups. We analyze these processes by focusing on those interfaces at which most of the energy or materials are exchanged.

10.4 Energy Flow

Primary productivity is apparently influenced by plant species richness only at levels far below those found in most mainland tropical forests (Wright, Chap. 2, this Vol.; Vitousek and Hooper 1993). Even highly fragmented and highly disturbed tropical forests have many more species than the minimum number needed to yield full primary productivity. Therefore, given that primary production capacity does not differ greatly among tropical forest trees, biomass production in tropical forests under relatively constant conditions probably is relatively insensitive to species richness.

However, fluctuations in rates of photosynthesis and biomass accumulation are strongly tied to the nature and intensity of disturbance. Because species differ in the speed with which they respond to disturbances, species richness may influence the rate at which biomass accumulates after disturbance (Denslow Chap. 7, this Vol.). In addition, variability in overall rates of photosynthesis per unit area may be inversely related to species richness if some species perform better in wet years and others in dry years; that is, richness may buffer primary production when environmental conditions vary (Tilman and Downing 1994). Variation in performance among tropical forest tree species is to be expected, but relevant data are yet to be gathered. Because tropical wet forest trees do not typically form annual growth rings, gathering data to measure the extent to which tree species richness buffers primary production when weather varies and perturbations strike may be difficult. New methods to estimate growth rates of tropical trees may make the task easier (Worbes and Junk 1989).

Forests that are naturally low in species richness are widely distributed in the tropics, often, but not always, on unusual soils (Connell and Lowman 1989; Hart and Murphy 1989; Hart 1990). However, these forests have not yet been studied sufficiently to determine the degree to which their productive capacity is influenced by low species richness, or to know if their interannual variation in total production is more sensitive to environmental variability than that of forests with greater species richness.

10.4.1 Carbon Allocation and Consumption

Secondary production, the summation of the growth of individuals and populations of all heterotrophic organisms, is completely dependent on primary production, nearly all of which in tropical forests results from photosynthesis by green plants. Relationships between primary production and secondary production are difficult to measure because many consumers are small or mobile and because traces of their consumption of plant tissues may disappear rapidly. However, it is evident that secondary production is not a simple function of primary production because plants have evolved a number of defensive structures and chemicals that deter consumption of their tissues by herbivores, parasites, and pathogens. These defenses also lower the digestive efficiency of consumers (McNaughton et al. 1989).

Plants influence animal biodiversity and productivity in two ways. They provide the energy that supports animal populations and they provide physical, temporal, and biochemical heterogeneity. Wood, roots, sap, floral and extrafloral nectar, leaves, flowers, fruits, and seeds differ strikingly in their physical and chemical structure, and the guilds of animal species that use those different tissues and fluids overlap relatively little.

Consumers may influence primary productivity by maintaining individual plants and plant populations in rapid growth phases by reducing the accumulation of living plant biomass, by reducing respiratory losses, and by recycling nutrients. These effects have been shown to be important in tropical savannahs (McNaughton 1985), but in tropical forests, where the amount of standing biomass is high relative to net primary production, the small amount of net primary production typically consumed by herbivores probably has little effect on total net primary production (Huston and Gilbert Chap. 3, this Vol.).

The quantity of secondary production and its distribution among species are both potentially influenced by species richness because different plants allocate their primary production in highly distinctive ways (Coley et al. 1985). Plant species differ in the proportion of primary production they allocate to defenses, which defensive compounds they synthesize, the quantities and composition of tissues that function to attract mutualists (Coley and Aide 1991; Davidson et al. 1991), and the physical and chemical composition of their wood. Because herbivore pressure is high throughout the year in moist forests, tropical woody plants allocate relatively large amounts of energy to chemical defenses (Levin 1978; Levin and York 1978; McKey 1979; Coley and Aide 1991) and resources that attract predators and parasites of herbivores (Simms 1992).

A higher percentage of primary production is consumed by herbivores in grasslands than in forests, However, if only production of foliage is considered, then percent consumption by herbivores is similar in the two ecosystem types (McNaughton et al. 1989). The available data suggest that secondary chemicals and physiological differences among herbivores have much less influence on average levels of herbivory at the ecosystem level than at individual and population levels. That is, defensive chemicals regulate what herbivores eat and the types of herbivores present more than they regulate overall levels of consumption.

"Mobile link" species (Gilbert 1980), such as pollinators and seed dispersal agents have little influence on fluxes of energy and materials in ecological time, but they may be critical to the maintenance of the species richness of tropical forests. Even though most frugivores are dietary generalists, many plants depend upon a small suite of frugivores for dispersing their seeds. Loss of those species is expected to have major influences on the long-term population dynamics of many tree species (Howe and Smallwood 1982; Terborgh 1986a). Maintenance of the frugivore functional group may depend upon the presence of a small number of tree species, e.g., Ficus spp. or Cecropia spp., that ripen their fruits at seasons of fruit scarcity (Terborgh 1986b).

10.4.2 Animal-Animal Interactions

The animals that eat the tissues of tropical forest plants support a rich assemblage of commensals, predators, parasites, and parasitoids, many of which, such as blood parasites of vertebrates, predatory mites, and parasitoid wasps, are tiny and inconspicuous. Though small, these organisms probably substantially reduce herbivory in both polyculture crop systems (Andow 1984) and natural vegetation (Gilbert 1977; Lawton and Brown 1993). Thus, by controlling herbivore populations, these tiny organisms may act as rate regulators or "energy filters" (Hubbell 1973), thereby reducing both the rate at which, and number of pathways by which, primary production becomes secondary production. For example, herbivorous insect larvae, which are major consumers of primary production, are attacked by both specialist and generalist invertebrate predators that typically maintain their populations well below outbreak levels (Huston and Gilbert, Chap. 3, this Vol.). Social wasps and ants, which are important predators of foliage-eating insects, are, in turn, attacked by army ants. Similarly, the intensity of grazing and browsing of tropical understory plants by vertebrates may be greatly reduced by predators. In forest fragments lacking these predators, browsing vertebrates can dramatically alter the structure and species composition of understory vegetation (Dirzo and Miranda 1991).

10.4.3 Detritus-Detritivores

Energy flow in ecosystems would quickly be reduced if the activities of detritivores were depressed or eliminated. Because rates of decomposition of fine detritus on the floors of tropical wet forests are very high and because in many (but not all – Jordan 1985, Brown and Lugo 1990) tropical forests most nutrients in the system are found in the bodies of plants, not in the soil, changes in rates of energy processing by soil detritivores exert a major influence on energy flows in those forests.

The taxonomy of tropical microbes is, unfortunately, extremely poorly known, and knowledge of the functional properties of even the named species is meager (Lodge et al., Chap. 5, this Vol.). Lack of relevant information prevents us from identifying useful functional groups of tropical forest microbes and knowing how many species can cleave particular chemical bonds in detritus. Determining the sensitivity of ecosystem processes to deletions of microbial species and which functional processes are likely to have the least functional redundancy will depend upon rectifying these knowledge deficiencies (Lodge et al., Chap. 5, this Vol.).

Tropical trees vary in tissue chemistry (Golley 1983a, b; Rodin and Basilevich 1967), in what they remove from the soil, and what they return to the soil. The meager evidence so far available suggests that trees of different species may generate significant differences in the soils in the area affected by their roots and litterfall. Soils under the legume *Pentaclethra macroloba* at La Selva, Costa Rica, have lower pH values than soils from areas away from individuals of this species, presumably because symbiotic microorganisms associated with *P. macroloba* trees fix nitrogen, which is then nitrified (Sollins, unpubl.). In contrast, Montagnini and Sancho (1994) did not observe a drop in pH attributable to nitrification in soils under two N-fixing legume species relative to that under non-N-fixing tree species in 4-year-old monospecific stands in the same area. This was so even though rates of net nitrification potential were higher under the legumes. Soils under female *Trophis involucrata* individuals have higher phosphorus concentrations than soils under males (Cox 1981). Incorporating N-fixing trees in production systems often increases yields of associated crops or trees (Alpizar et al. 1986; Dommengues 1987; Szott et al. 1991), but whether soil differences generated by trees in natural forests influence regeneration, growth, and species richness in tropical forests remains to be determined (Parker 1994). However, there is evidence to suggest that such effects can be expected and that they can result in effects on ecosystem functioning. For example, the tree *Myrica faya* and its associated N-fixing bacteria, significantly affect soil nitrogen, productivity, and mineral cycling on Hawaiian laval flows (Vitousek and Walker 1989).

10.5 Materials Processing

The movement of materials in ecosystems is often strongly tied to movement of energy, but the two processes are disconnected often enough to
favor considering materials processing separately from energy flow.

10.5.1 Atmosphere-Organism

Plants are influenced by two counterflowing streams of water. Water moves
within plants from soil into roots, upward through stems, and is transpired
from leaves. Water deposited on plant surfaces by precipitation drips
through branches and falls from foliage (throughfall) or runs down stems
to the soil (stemflow). These flows influence, and are influenced by, the
plants themselves, and they alter conditions for all other organisms living
in and around the plants.

Interception of fog and mist is influenced by the size and orientation of
leaves and thalli of macrolichens. The existence of low stature evergreen
wet forests on the summits of otherwise dry mountains adjacent to the
Caribbean coast of South America is possible only because fog interception
accounts for up to two-thirds of the monthly water input to these forests
(Cavelier and Goldstein 1989).

Plants, photosynthetic microorganisms, and nitrogen fixers also are
actively and massively involved with direct exchanges of materials with the
atmosphere, but measurements of air quality above and within tropical
rain forests are rare and atmosphere-canopy exchange of aerosols in tropical forests is poorly understood. Plants intercept airborne particles, either
as dry or wet deposition, and release to the atmosphere carbon dioxide, a
variety of volatile organic compounds, and large quantities of water.
Animals and microbes also release large quantities of carbon dioxide and
methane to the atmosphere. Exchanges of these materials by both groups
appear to be directly proportional to the total biomass of organisms irrespective of its distribution among species, except that exchange of materials is lower in systems dominated by woody plants.

Epiphytes depend primarily on direct nutrient exchange with the atmosphere for their nutrients and water. Species richness of epiphytes is high in
most tropical forests (Benzing 1990; Gentry and Dodson 1987), and because
they tend to grow on different parts of the trees, nutrient exchange rates
may depend upon both the richness of epiphyte species and their biomass
(Silver et al., Chap. 4, this Vol.). The best evidence for the role of epiphytes
in nutrient cycling comes from tropical cloud forests, where nearly half of
the foliage nutrient pool may be stored in epiphyte biomass (Nadkarni
1984) and nutrient availability is often low because low temperatures and
high moisture retard decomposition. Also, because the volatile organic
compounds produced by plants are highly species-specific (Fall 1991), the

composition of airborne volatiles may carry a signature of the species richness of the forest canopy.

Unlike most other nutrients, the major sources of nitrogen to ecosystems are precipitation and biological nitrogen-fixation by free-living bacteria and cyanobacteria (Bentley 1987; Bentley and Carpenter 1984), by bacteria having mutualistic associations with plants and fungi, and by gut-dwelling symbionts of termites (Prestwich et al. 1980, Prestwich and Bentley 1981). Nitrogen-fixing organisms, especially cyanobacteria, fix substantial amounts of nitrogen relative to other sources in some tropical forests (Lodge et al. Chap. 5, this Vol.). In species-poor systems, such as those growing on young tropical lava flows, invasion of a single tree species and lichenized fungi with nitrogen-fixing bacteria may dramatically increase nitrogen input to the system, productivity, and ecosystem development (Vitousek and Sanford 1987; Vitousek and Walker 1989).

10.5.2 Biotic Interface

In tropical forests large quantities of nutrients are stored in live biomass (Jordan 1985). The synthesis of defensive chemicals, by reducing losses to herbivores, may result in conservation of nutrients (McKey et al. 1978; Hobbie 1992), especially in forests where leaves are long lived. Within-plant nutrient transfer may be significant for conservation of some nutrients, particularly phosphorus (Vitousek 1984). Interspecific differences in tissue chemistry, combined with differences in the degree to which plants recapture nutrients prior to discarding their leaves, could result in nutrient dynamics being influenced by tree species richness. Studies to test this possibility have not yet been carried out.

10.5.3 Plant-Soil

Nutrients are usually taken up from the soil and forest floor by mycorrhizal fungi associated with fine roots and eventually returned again through decomposition of litterfall and belowground litter inputs (Went and Stark 1968; Stark 1971; Stark and Jordan 1978; Cuevas and Medina 1988). The decomposition of litter is carried out primarily by microbes and soil-dwelling animals whose diversity and functioning are still poorly known. The role of microorganisms in the nitrogen cycle appears to be especially important for productivity and biomass accumulation in tropical forests because different plant species require nitrogen in different inorganic forms (NH_4, NO_2, NO_3). An important functional group of organisms in tropical forests are endomycorrhizal fungi that have mutualistic relationships with at least 80 % of tropical plants (Janos 1983). The growth of tropical moist forest trees may be especially sensitive to losses in microbial diversity because, unlike temperate forests, which have relatively few tree species but many fungal species, tropical forests have many tree species but relatively few species of endomycorrhizal fungi (Malloch et al. 1980). Furthermore, these fungi may differ significantly in the benefits they offer to different host species (Miller 1989). Large-scale, long-term conversion of forests to grasslands or cropland results in major changes in soil nutrient pools and the soil biota (Olson 1963; Hamilton and King 1983; Macedo and Anderson 1993; Henrot and Robertson1994) which reduces nutrient cycling rates in those agro-ecosystems and decreases the potential for regenerating forests on those lands.

10.5.4 Atmosphere-Soil

Soil microorganisms release to the atmosphere large quantities of carbon dioxide and various nitrogen compounds (Yoshinari 1993), and the soil surface receives atmospheric deposition as throughfall and stem flow. Currently tropical forests are a net source of atmospheric CO_2, but this is due to reduction of total area of forests and to extensive burning, not to loss of species per se (Detwiler et al. 1988; Hall and Uhlig 1991; Houghton 1991). Tropical forests release large quantities of methane to the atmosphere, much of it due to the activities of methanogenic bacteria in tropical wetlands (Bartlett and Harriss 1993). Gut symbionts of termites are also a significant source of methane. The influence of biodiversity on rates of emission and consumption of methane and other chemicals is unknown, and the lack of information on the identities and functional attributes of most soil microorganisms makes it impossible to identify significant functional groups and the number of species found in most of those groups.

10.5.5 Soil-Water Table

Most nutrients leave forested ecosystems through the soil. Tropical forests growing on deep, highly weathered soils have low nutrient-holding capacities (Sánchez 1976), and the large volumes of water that move through the soil generate high potential losses of nutrients through leaching (Radulovich and Sollins 1991). Rates of movement of water on and within the soil are reduced by woody debris and fine litter and by the extensive mats of fine roots that characterize tropical forests, especially those growing on nutrient-poor soils (Silver et al. Chap. 4, this Vol.). Plant roots also recapture nutrients in the soil, and deep-rooted species may pump water from deep in the soil to the surface where it may be transpired. Tropical trees differ in the depth to which their roots penetrate, with the result that the rate of loss of nutrients via leaching may be inversely related to species richness, but no data exist to test these relationships quantatively.

10.6 Functional Properties Over Longer Temporal Scales

Until now, we have concentrated on processes occurring at a specific location and on scales of days to a few years. However, the functioning of tropical forest ecosystems depends on the formation and maintenance of structure, which is the result of photosynthesis and biomass accumulation and
biogeochemical cycling over many decades or centuries. The ways in which
forests respond to perturbations, such as fire, drought, strong winds,
unusually heavy rains, invasions of exotic species, and losses of species
typically found in the system, are also important, and these responses may
be influenced by plant species richness in ways that are not apparent under
constant conditions.

10.6.1 Provision and Maintenance of Structure

The many plant species that live in tropical moist forests can be grouped
into a small number of life forms. Most species of canopy trees are similar
enough in their growth forms that loss of particular species would not
change vegetation structure very much (Ewel and Bigelow, Chap. 6, this
Vol.). However, certain life forms, particularly palms, lianas, and epiphytic
bromeliads, are both structurally highly distinctive and represented by
relatively few species. Absence of these structural elements may influence
the degree to which the forests are altered by perturbations and their rate
of recovery after perturbations (Denslow, Chap. 7, this Vol.).

10.6.2 Resistance to Invasions

Many exotic species have been introduced into tropical regions, as they
have into temperate regions (Drake 1989). However, existing data suggest
that exotic species seldom have invaded undisturbed mainland tropical
forests (Ramakrishnan 1991). Weedy plants and animals are confined
primarily to highly disturbed habitats. The high rate of recovery of wet
tropical vegetation after natural and human-made disturbances may be the
key factor or, at least, part of the story (Rejmanek, Chap. 8, this Vol.). Both
abiotic (high moisture, radiation, and temperature) and biotic (pool of
immediately germinating, fast growing, already naturalized species) components accelerate recovery rates. Few invaders can survive in the presence
of robust and rapidly growing pioneer species. The scarcity of such species
on islands may be partly responsible for the vulnerability of island forests
to plant invasions. Mature tropical rain forests can be invaded by only few
(mostly shade-tolerant) plant species. More than 60% of woody species
invading primary tropical rain forests are dispersed by native or introduced vertebrates (Rejmanek, Chap. 8, this Vol.), This is consistent with the

fact that usually more than 60% of woody species in tropical forests are adapted for seed dispersal by vertebrates. The same vertebrate species responsible for maintenance of diversity in tropical forests may also be responsible for their invasion by exotic species.

10.7 Functional Properties Over Larger Spatial Scales

The approach explored in this book, although it has focused on local processes and interactions, may help us understand how species richness influences large-scale processes. Linkages across space are the result of movement of materials across landscapes by various physical and biological agents. Effects on ecosystem processes are functions of the distances over which materials move. Here we explore how those transfers may be influenced by species richnesss (Table 10.1).

Table 10.1. Landscape scale linkages in tropical forests

Linkage	Process	Distance moved	Distance effect	Influence of species richness
Atmosphere-organism	Release of CO_2 CH_4, volatiles, transpiration	Local – regional	Atmospheric deposition	Probably none
		Local – global	Increased precipitation	Probably none
Atmosphere-soil	Release of CO_2, CH_4, volatiles, evaporation	Local – regional	Atmospheric	Probably none
		Local – global	Increased precipitation	Probably none
Soil - water	Leaching	Local – watershed	Riparian deposition	Dependent on efficiency of root mat
	Movement of H_2O to water table	Local – watershed	Recharge of groundwater	Dependent on density of phreatophytes
Animal movement	Seasonal migration	Latitudinal– global	Spread of disease and propagules	Proportional to number of migrants
	Within-tropics migration,	Local – regional	"	"
	breeding concentrations (birds, ants termites)	Local – regional	Nutrient concentration, local resource depletion, patchy deposition of seeds	Dependent on presence of colonial species

10.7.1 Movement of Materials by Physical Agents

As indicated in Table 10.1, many linkages that connect local tropical forest ecosystems across space are strongly influenced by the total amount of vegetative cover, but they are probably little influenced by species richness per se. The variety of tree species may well influence the composition of the airborne volatile organic compounds, but the significance of this relationship has not been determined.

Tropical forests are currently a net source of atmospheric carbon dioxide, due to all forms of oxidation, much of it resulting from a reduction in total acreage and subsequent burning. The sequestering of carbon by growing tropical forests is apparently poorly correlated with plant species richness because tropical plantations can and do accumulate carbon at rates similar to or greater than those of natural species-rich forests of the same age (Cuevas et al. 1991; Ewel et al. 1991; Lugo 1992).

Many tropical rivers and streams flowing through inhabited areas are polluted. Species richness is reduced in polluted rivers, as found in lead- and mercury-polluted segments of the Orinoco River as a result of gold mining (Pfeiffer and De Lacerda 1988), but there are no theoretical or empirical reasons to believe that changes in water quality are related to loss of species. Similarly, the quantities of water flowing and the temporal pattern of flows is strongly influenced by vegetative cover, particularly the loss of forests, but the influence of plant species richness per se for patterns of water discharge from tropical forests is also probably small.

Deforestation is dramatically altering tropical forest landscapes and waterscapes. Such fragmentation is evidently leading to loss of species, but how loss of species may, in turn, influence the structure of tropical landscapes or waterscapes is unclear, and there are currently no theories that predict such relationships.

10.7.2 Movement of Materials and Energy by Animals

The transfer of most energy across tropical landscapes is the result of movement of animals. Tropical regions are invaded each year by many thousands of migrant birds that breed at high latitudes but winter in the tropics. Migrants may outnumber residents during part of the year in some tropical habitats. These migrants may compete with themselves and with residents for food (Keast and Morton 1980; Greenberg 1986) and they are potential sources of disease transmission, although little is known about the diseases of tropical birds or whether migrants are sources of infections in resident species.

Many species of birds, butterflies, and moths migrate seasonally within the tropics, either elevationally (especially nectarivorous and frugivorous species) or from dry to wet forests during dry seasons (Stiles 1988; Loiselle

1991). These migrants are also potential movers of pathogens and they carry large numbers of propagules across the landscape. The importance of migration corridors and suitable areas in which to live throughout the year are important for the viablity of populations of within-tropical migratory species, but the consequences of the potential loss of those species for the functioning of tropical forests are yet to be investigated.

Animals may have a variety of effects on nutrient processing in tropical forests. Animals mix soils and promote aggregate soil structure, redistribute and concentrate canopy tissues in the soil, release methane (wood-eating insects having methane-producing gut symbionts), and produce readily decomposed frass and feces. Moving animals spread diseases and plant propagules and concentrate nutrients around their nests, roosts, and resting places. Nutrient "hot spots," such as termite mounds (Nye 1955; Cox and Gakahu 1985; Oliveira-Filho 1992), may be important for regeneration of trees and other processes. In Neotropical forests, leaf-cutting ants concentrate large quantities of nutrients in and around their large subterranean nests (Haines 1975). Within tropical forests most of these processes are influenced by many species of animals, but because typically one or a few species of leaf-cutting ants harvest most of the leaves in a particular forest, nutrient-concentrating processes may be highly sensitive to the loss of a single species.

10.8 Biodiversity and Responses to Disturbances

Human activities cause a diverse array of disturbances to natural ecosystems, among which are increasing levels of atmospheric CO_2, changing climates, increasing fragmentation of habitats, and introductions of species into regions where they were previously absent. Many of these changes cause extinctions of species. Given these and other large-scale changes to tropical and nontropical systems, it is important to determine whether or not there are links between the components of biodiversity and the ability of ecosystems to withstand and recover from such alterations. As touched on by Silver et al. (Chap. 4) and Denslow (Chap. 7, both this Vol.), such links may exist with respect to redundancy within functional groups. Taxa within the same functional group, although they affect common processes, may differ in their responses to human perturbations. Groups of species having similar ecosystem effects, but which differ in responses to perturbations, may buffer ecosystem processes and products in the face of present and future human-caused perturbations of ecological systems.

10.9 Research Agenda

In this book we have concentrated our attention on relationships between biological diversity and ecosystem processes in tropical moist forests, highlighting the shortage of information that is needed to assess biodiversity-ecosystem functional relationships. The shortage is even more dramatic for other types of tropical forests, which have not received the attention that has been directed toward wet forests. Dry forests, montane forests, and wetlands, because of the magnitude of anthropogenic alteration they are experiencing and their crucial contribution to tropical biological diversity, must be better understood if we are to document the range of variation of the relationships between biological diversity and functioning of tropical ecosystems. Interestingly, changes in patterns and processes are similar along gradients of temperature, fertility, and moisture in spite of the dramatically different environmental changes associated with each one. Thus, decreases in productivity, standing biomass, decomposition rates, life form diversity, and species richness, and increases in belowground allocation of plant resources, accompany all three gradients. Determining the reasons for similar responses to these different environmental conditions is a challenge for future tropical forest research.

Ecologists traditionally evaluate primary production by aggregating into one measure the total amount of carbon fixed per unit area. Although such estimates are useful for some purposes, these data have a number of significant limitations. Lumped values underestimate aboveground primary production because they ignore carbon allocated to nectar, flowers, and fruits – key resources for a range of influential consumer groups – and allocation downward into roots and mycorrhizae. Lumped primary production tells us nothing about how carbon is allocated among different plant parts (roots, wood, leaves, nectar, flowers, and fruits); in essence primary producers are aggregated into a single functional group. Plant materials are packaged in fundamentally different ways, and these differences determine the identity of consumer groups and the rates of consumption and energy flow in ecosystems. To assess the contribution of different species, functional groups, and life-forms to ecosystem level processes, we need disaggregated estimates of primary production.

One of the most important areas of research on relationships between biodiversity and ecosystem functioning concerns the influence of the species composition of leaf litter on rates of decomposition and subsequent mineralization. Limited data suggest that decomposition rates can be influenced by litter diversity; rates are higher for litter from richer species assemblages than from poorer species assemblages (Burghouts et al. 1994). The implications of this relationship for tropical systems are profound because decomposition is such a pivotal process.

Remarkably little is known about belowground plant and microbial processes. Remedying this situation needs to be a major priority if we are to understand the links between biodiversity and ecosystem-level processes. Three areas are in particular need of attention. First, are there predictable structural patterns in the root systems of multispecies tropical assemblages and is structural complexity correlated with species richness? Second, how species-rich is the soil microbiota and into how many functional groups do these species fall? Third, when tropical forests are perturbed such that species or groups of species are deleted, do compensatory responses by other taxa result in reoccupation of the space and restoration of ecological processes?

The research effort requires both manipulative experiments and comparative studies that contrast (1) naturally occurring mono-specific or low diversity forests with neighboring high diversity forests, (2) human-damaged systems with neighboring undisturbed systems, (3) CO_2-enriched systems with systems under current ambient CO_2 concentrations, and (4) forests along gradients of precipitation, soil fertility, and elevation. Only with such studies can a comprehensive picture of the importance of biodiversity for tropical forest functioning be developed.

Sampling protocols for a number of ecological processes are typically carried out at weekly, monthly, or bimonthly rates. For some issues relevant to functioning of tropical forests, however, sampling at more frequent intervals is necessary. This is particularly evident in the case of the sudden pulses of nutrient capture and transfer of nutrients which occur after a quick wetting of the soil. Such sudden pulses determine patterns at the Materials Processing Interface, and they would go unnoticed unless very frequent sampling is done.

10.10 Conclusions

Humanity needs to protect and nourish tropical forests for many reasons. Among them are components that derive from the biological complexity of tropical forests. Because of their complexity, tropical forests have an extremely high information content that resides in the genomes of the individual species, interactions among them, and the resulting ecosystem processes. Most of this information is not yet accessible to us because we have described only a small fraction of the species living in tropical forests, we know almost nothing about ecological relationships among the species we have described, and we have measures of the rates and magnitudes of some ecological processes at only a few tropical forest sites.

Many benefits can be derived from preserving and studying tropical forests. With better knowledge of the players and the theater we will gain a better understanding of how complex systems work. To live sustainably on

Earth, humans will need to manipulate the dynamics of many kinds of complex physical, biological, and social systems. Many degraded environments, both tropical and temperate, need to be restored. Knowledge of how tropical forests work is certain to be helpful in design, development, and execution of restoration efforts worldwide. Increasingly, humans are required to manage ecosystems more intensively to increase production of desired products, reduce losses of energy and materials through undesired channels, and to establish integrated landscapes whose components interact in ways that improve the rates and stabilities of processes that maintain those systems. Management plans are more likely to achieve their objectives if they are based on solid understanding of the behavior of the systems being managed rather than on untested guesses.

Throughout its history, humankind has drawn upon tropical forests for products that enrich human life. Important among these are food, fiber, medicines, drugs, and esthetic pleasure. All of these goods and services scale directly on biological diversity of the forests. The woods and fibers of different species are useful for different purposes. The chemicals synthesized by living organisms that are the basis of medicines and drugs tend to be highly species-specific (sometimes they are strain-specific) or at least confined to a small number of species, usually closely related ones. Options to find and use new products are sacrificed when forests are lost and biological diversity is reduced. As a result of our poor understanding of how tropical forests work, we may inadvertently cause losses of many species living in those forests we do preserve. In so doing we further reduce options and make the remaining forests ever more vulnerable to perturbations they can currently withstand.

Perhaps the feature of tropical forests that most hinders our ability to understand their dynamics is the slow rate at which they change over time. The magnificent trees that dominate and give structure to tropical forests live, on average, more than a century. Some live much longer. Once a tree has gained its position in the forest canopy, it usually survives until long after the clues about the causes for its initial success have totally disappeared. Its current associates may be quite different from those it had when it was young; the local climate may have changed as well. Only about 50 tree generations have elapsed since the final retreat of the last Pleistocene glaciers. During glacial advances, temperatures dropped on average about 6 °C in tropical lowlands (Bush and Colinvaux 1990). Pollen profiles from tropical regions show that trees now restricted to middle elevations on mountainsides were intermingled with lowland trees close to sea level (Bush et al. 1990; Colinvaux et al. 1996). During glacial maxima, levels of atmosperic CO_2 were much lower than they were 100 years ago and still lower than that relative to today's levels. Tropical forests are probably still readjusting to postglacial climatic changes. Some types of disturbances, such as fires and hurricanes, produce immediate and sometimes catastrophic effects on tropical forests and their functional properties (Boucher et al. 1990; Walker et al.1991). The longevity of trees causes long lags between the imposition of some types of disturbances and measurable responses in forest functioning. For example, loss of specific frugivores may not affect composition and functioning of a forest for more than a century, even if that loss eventually results in the extirpation of a suite of forest tree species.

For these reasons, much attention should be given to understanding the rates at which different perturbations are likely to affect tropical forest functioning, which processes they affect, how they exert their influences, and the time frames over which their affects are likely to be expressed. The processes that unfold slowly are the ones most likely to be ignored and unappreciated, yet they may ultimately be among the most important determinants of the long-term functioning of tropical forests.

References

Aide TM (1988) Herbivory as a selective agent on the timing of leaf production in a tropical understory community. Nature 336:574-575

Alpizar L, Fassbender HW, Heuveldop J, Folser H, Enriquez G (1986) Modelling agroforestry systems of cacao (*Theobroma cacao*) with laurel (*Cordia alliodora*) and poro (*Erythrina poeppigiana*) in Costa Rica. I. Inventory of organic matter and nutrients. Agrofor Syst 4:175-189

Andow D (1984) Effects of agricultural diversity on insect populations. In: Lockeretz W (ed) Environmentally sound agriculture. Prager, New York, pp 96-115

Baker HW, Bawa KS, Frankie GW, Opler PA (1983) Reproductive biology of plants in tropical forests. In: Golley FB (ed) Tropical rain forest ecosystems: structure and function. Elsevier, Amsterdam, pp 183-215

Bartlett KB, Harriss RC (1993) Review and assessment of methane emission from wetlands. Chemisphere 26:261-320

Bawa KS (1979) Breeding systems of trees in a tropical wet forest. NZ J Bot 17:521-524

Bawa KS (1990) Plant-pollination interactions in tropical lowland rain forest. Annu Rev Ecol Syst 21:254-274

Bawa KS, Beach JH. (1981) Evolution of sexual sytems in flowering plants. Ann Mo Bot Gard 68:254-274

Bawa KS, Hadley M (eds) (1990) Reproductive ecology of tropical forest plants. UNESCO, Parthenon

Beaver RA (1979) Host specificity of temperate and tropical animals. Nature 281:139-141

Bentley BL (1987) Nitrogen fixation by epiphylls in a tropical rainforest. Ann Mo Bot Gard 74:234-241

Bentley BL, Carpenter EJ (1980) Effects of desiccation and rehydration on nitrogen fixation by epiphylls in a tropical rainforest. Microbiol Ecol 6:109-113

Bentley BL, Carpenter EJ (1984) Direct transfer of newly-fixed nitrogen from free-living epiphyllous microorganisms to their host plant. Oecologia 63:52-56

Benzing DH (1990) Vascular epiphytes. Cambridge Univ Press, Cambridge

Birch HF (1958) The effect of soil drying on humus decomposition and nitrogen availability. Plant Soil 10:9-13

Boucher DH, Vandermeer JH, Yih K, Zamora N (1990) Contrasting hurricane damage in tropical rain forest and pine forest. Ecology 71:2022-2024

Braithwaite RW (1987) Effects of fire regimes on lizards in the wet-dry tropics of Australia. J Trop Ecol 4:77-88

Brandani A, Hartshorn GS, Orians GH (1988) Internal heterogeneity of gaps and species richness in Costa Rican tropical wet forest. J Trop Ecol 4:99-119

Brokaw NVL (1982) Treefalls: frequency, timing and consequences. In: Leigh EG, Rand AS, Windsor DM (eds) The ecology of a tropical forest. Seasonal rhythms and long-term changes. Smithsonian Press, Washington DC

Brown JH (1981) Two decades of homage to Santa Rosalia: toward a general theory of diversity. Am Zool 21:877-888

Brown S, Lugo AE (1990) Tropical secondary forests. J Trop Ecol 6:1-32

Brown S, Lugo AE, Silander S, Liegel L (1983) Research history and opportunities in the Luquillo Experimental Forest. Inst Trop For, USDA Gen Tech Rep SO-44

Brunig EF (1983) Vegetation structure and growth. In: Golley FB (ed) Ecosystems of the world 14A. Tropical rain forest ecosystems. Elsevier, Amsterdam, pp 49-75

Bullock SH, Solís-Magallanes JA (1990) Phenology of canopy trees of a tropical deciduous forest in Mexico. Biotropica 22:22-35

Burghouts TBA, Campbell EJF, Kolderman PJ (1994) Effects of tree species heterogeneity on leaf fall in primary and logged dipterocarp forest in the Ulu Segama Forest Reserve, Sabah, Malaysia. J Trop Ecol 101 -26

Bush MB, Colinvaux PA, Wiemann MC, Piperno DR, Liu K-b (1990) Late Pleistocene temperature depression and vegetation in Ecuadorian Amazonia. Quat Res 34:330-345

Bush MB, Colinvaux PA (1990) A long record of climatic and vegetation change in lowland Panama. Vegetat Sci 1:105-118

Cavelier J, Goldstein G (1989) Mist and fog interception in elfin cloud forests in Colombia and Venezuela. J Trop Ecol 5:309-322

Coley PD, Aide TM (1991) Comparison of herbivory and plant defenses in temperate and tropical broad-leaved forests. In: Price PW, Lewinsohn TM, Fernandes GW, Benson WW (eds) Plant-animal interactions. Wiley, New York, pp 25-50

Coley PD, Bryant JP, Chapin FS (1985) Resource availability and plant-antiherbivore defense. Science 230:895-899

Colinvaux PA, Liu K-b, DeOliveira P, Bush MB, Miller MC, Steinitz Kannan M (1996) Temperature depression in the lowland tropics in glacial times. Climate Change 32: 19-33

Connell JH, Lowman MD (1989) Low-diversity tropical rain forests: Some possible mechanisms for their existence. Am Nat 134:88-119

Cox GW, Gakahu CG (1985) Mima-mound micro-topography and vegetation patterns in Kenyan savannas. J Trop Ecol. 1:23-26

Cox PA (1981) Niche partitioning between sexes of dioecious plants. Am Nat 117:295-307

Cuevas E, Medina E (1988) Nutrient dynamics within Amazonian forests. II. Fine root growth, nutrient availability and leaf-litter decomposition. Oecologia 76:222-235

Cuevas E, Brown S, Lugo AE (1991) Above- and belowground organic matter storage and production in a tropical pine plantation and a paired broadleaf secondary forest. Plant Soil 135:257-268

Currie DJ, Paquin V (1987) Large scale biogeographical patterns of species richness patterns of trees. Nature 329:326-327

Davidson DW, Foster RB, Snelling RR, Lozada PW (1991) Variable composition of tropical ant-plant symbioses. In: Price PW, Lewinsohn TM, Fernandes GW, Benson WW (eds) Plant-animal Interactions: evolutionary ecology in tropical and temperate regions. Wiley, New York, pp 145-163

Dirzo R, Miranda A (1991) Altered patterns of herbivory and diversity in the forest understory: a case study of the possible consequences of contemporary defaunation. In: Price PW, Lewinsohn TM, Fernandes GW, Benson WW (eds) Plant-animal interactions. Wiley, New York, pp 273-288

Dommengues YR (1987) The role of biological fixation in agroforestry. In: Steppler HA, Nair PKR (eds) Agroforestry. A decade of development. Int Counc Res Agrofor, Nairobi, pp 245-271

Drake JA (ed) (1989) Biological invasions: a global perspective. Wiley, Chichester

Eguiarte LE, Búrquez A, Rodriguez J, Martínez-Ramos M, Sarukhán J, Pinero D (1993) Direct and indirect estimates of neighborhood and effective population size in a tropical palm, *Astrocaryum mexicanum*. Evolution 47:75-87

Estrada A, Fleming TH (eds) (1986) Frugivores and seed dispersal. Junk, Dordrecht, 392 pp

Ewel JJ, Mazzarino MJ, Berish CW (1991) Tropical soil fertility changes under monocultures and successional communities of different structure. Ecol Appl 1:289-302

Fall R (1991) Isoprene emission from plants: summary and discussion. In: Sharkey TD, Holland EA, Mooney HA (eds) Trace gas emissions by plants. Academic Press, New York, pp 209-215

Food and Agricultural Organization (1993) Tropical forest resources assessment project. FAO, Rome

Foster RB (1982a) The seasonal rhythm of fruit fall on Barro Colorado Island. In: Leigh E, Rand AS, Windsor DM (eds) The ecology of a tropical forest. Seasonal rhythms and long-term changes. Smithsonian Press, Washington DC, pp 151-172

Foster RB (1982b) Famine on Barro Colorado Island. In: Leigh EL, Rand AS, Windsor DM (eds) The ecology of a tropical forest. Seasonal rhythms and long-term changes. Smithsonian Press, Washington DC, pp 201-212

Gentry AH (1982) Patterns of neotropical species diversity. Evol Biol 15:1-84

Gentry AH (1988) Changes in plant community diversity and floristic composition of environmental and geographical gradients. Ann Mo Bot Gard 75:1-34

Gentry AH (ed) (1990) Four neotropical forests. Yale Univ Press, New Haven

Gentry AH (1992) Tropical forest biodiversity: distributional patterns and their conservational significance. Oikos 63:19-28

Gentry AH, Dodson C (1987) Contribution of nontrees to species richness of a tropical rain forest. Biotropica 19:149-156

Gentry AH, Emmons LH (1987) Geographic variation in fertility, phenology, and composition of the understory of Neotropical forests. Biotropica 19:217-227

Gilbert LE (1977) The role of insect-plant coevolution in the organization of ecosystems. In: Labyrie V (ed) Comportement des insectes et milieu trophique. CNRS, Paris, pp 399-413

Gilbert LE (1980) Food web organization and conservation of neotropical diversity. In: Soulé ME, Wilcox B (eds) Conservation biology. Sinauer, Sunderland, MA, pp 11-33

Gilbert LE, Smiley JT (1978) Determinants of local diversity in phytophagous insects: host specialists in tropical environments. In: Mound LA, Waloff N (eds) Diversity of insect faunas. Symp R Entomol Soc Lond 9:89-105

Golley FB (1983a) The abundance of energy and chemical elements. In: Golley FB (ed) Ecosystems of the world, vol 14A. Tropical rain forest ecosystems. Elsevier, Amsterdam, pp 101-115

Golley FB (1983b) Nutrient cycling and nutrient conservation. In: Golley FB (ed) Ecosystems of the world, vol 14A. Tropical rain forest ecosystems. Elsevier, Amsterdam, pp 137-156

Green GM, Sussman RW (1990) Deforestation history of the eastern rain forests of Madagascar from satellite images. Science 248:212-215

Greenberg R (1986) Competition in migrant birds in the nonbreeding season. Curr Ornithol 3:281-307

Grubb PJ (1977a) Control of forest growth and distribution on wet tropical mountains: with special reference to mineral nutrition. Annu Rev Ecol System 8:83-107

Grubb PJ (1977b) The maintenance of species richness in plant communities: the importance of the regeneration niche. Biol Rev 523:107-145

Haines BL (1975) Impact of leaf-cutting ants on vegetation development on Barro Colorado Island. In: Golley FB, Medina E (eds) Tropical ecological ecosystems. Springer, Berlin Heidelberg New York

Hall CAS, Uhlig J (1991) Refining estimates of carbon released from tropical land-use change. Can J For Res 21:118-131

Hamilton LS, King PN (1983) Tropical forest watersheds: hydrologic and soil resources to major users or conversions. Westview Press, Boulder

Hamrick JL, Loveless MD (1989) The genetic structure of tropical tree populations: associations with reproductive biology. In: Bock JH, Linhart YB (eds) The evolutionary ecology of plants. Westview Press, Boulder, pp 129-146

Hart TB (1990) Monospecific dominance in tropical rain forests. Trends Ecol Evol 5:6-11

Hart JA, Murphy PG (1989) Monodominant and species-rich forests of the humid tropics: causes for their co-occurrence. Am Nat 133:613-633

Henrot J, Robertson GP (1994) Vegetation removal in two soils of the humid tropics: effect on microbial biomass. Soil Biol Biochem 26:111-116

Hobbie SE (1992) Effects of plant species on nutrient cycling. Trends Ecol Evol 7:336-339

Houghton RA (1991) Tropical deforestation and atmospheric carbon dioxide. Climate Change 19:99-118

Howe H, Smallwood J (1982) Ecology of seed dispersal. Annu Rev Ecol Syst 13:201-228

Hubbell SP (1973) Populations and simple food webs as energy filters. I. One species systems. Am Nat 107:194-201

Huston M (1994) Biological diversity, soils and economics. Science 262:1676-1680

Intergovernmental Panel on Climate Change (1990) Climate change. The IPCC scientific assessment. Houghton JT, Jenkins GJ, Ephraums JJ (eds) Cambridge Univ Press, New York

Janos DP (1983) Tropical mycorrhizae, nutrient cycles, and plant growth. In: Sutton S, Whitmore TC, Chadwick AC (eds) Tropical rain forests: ecology and management. Blackwell, Oxford, pp 327-345

Janzen DH (1973) Comments on host-specificity of tropical herbivores and its relevance to species richness. In: Heywood V (ed) Taxonomy and ecology. Academic Press, New York, pp 201-211

Janzen DH (1980) Specificity of seed-attacking beetles in a Costa Rican deciduous forest. J Ecol 68:929-952

Janzen DH (1987) Insect diversity of a Costa Rican dry forest: why keep it, and how? Biol J Linn Soc. 30:343-356

Janzen DH, Schoener TW (1968) Differences in insect abundance between wetter and drier sites during a tropical dry season. Ecology 49:96-110

Jordan CF (1985) Nutrient cycling in tropical forest ecosystems. Wiley, New York

Jordan CF (1991) Productivity of a tropical forest and its relation to a world pattern of energy storage. J Ecol 59:127-142

Karr JR, Freemark KE (1983) Habitat selection and environmental gradients: dynamics in the "stable tropics." Ecology 64:1481-1494

Keast A, Morton ES (eds) (1980) Migrant birds in the Neotropics: ecology, behavior, distribution and conservation. Smithsonian Press, Washington DC

Lawton JH, Brown VK (1993) Redundancy in ecosystems. In: Schulze E-D, Mooney HA (eds) Biodiversity and ecosystem function. Springer, Berlin Heidelberg New York, pp 255-270

Lawton RO, Putz FE (1988) Natural disturbance and gap-phase regeneration in a wind-exposed tropical cloud forest. Ecology 69:764-777

Leigh EG (1975) Structure and climate in tropical rain forest. Annu Rev Ecol Syst 6:67-86

Leigh EG, Rand AS, Windsor DM (eds) (1990) Ecología de un bosque tropical: ciclos estacionales y cambios a largo plazo. Smithson Trop Res Inst, Panama, 546 pp

Leighton M, Leighton DR (1983) Vertebrate responses to fruiting seasonality within a Bornean rain forest. In: Sutton SL, Whitmore TC, Chadwick AC (eds) Tropical rain forest: ecology and management. Br Ecol Soc Spec Pub 2. Blackwell, Oxford, pp 181-196

Levey DJ, Moermond TC, Denslow JS (1993) Frugivory: an overview. In: McDade LA, Bawa KS, Hespenheide HA, and Hartshorn GS (eds) La Selva. Ecology and natural history of a Neotropical rain forest. Univ Chicago Press, Chicago, pp 282-294

Levin DA (1978) Alkaloids and geography. Am Nat 112:1133-1134

Levin DA, York BM (1978) The toxicity of plant alkaloids: an ecogeographic perspective. Biochem Syst Ecol 6:61-76

Lieberman D (1982) Seasonality and phenology in a dry tropical forest in Ghana. J Ecol 70: 791-806

Lodge DJ, McDowell WH, McSweeney CP (1994) The importance of nutrient pulses in tropical forests. Trends Ecol Evol 9:384-387

Loiselle BA (1991) Temporal variation in birds and fruits along an elevational gradient in Costa Rica. Ecology 72:180-193

Loveless MD, Hamrick JL (1987) Distribución de la variación genética en especies arboreas tropicales. Rev Biol Trop 35:165-175

Lugo AE (1988) Estimating reductions in the diversity of tropical forest species. In: Wilson EO, Peter FM (eds) Biodiversity. Nat Acad Press, Washington DC, pp 58-70

Lugo AE (1992) Comparison of tropical tree plantations with secondary forests of similar age. Ecol Monogr 62:1-41

Lugo AE, Parrotta J, Brown S (1993) Loss in species caused by tropical deforestation and their recovery through management. Ambio 22:106-109

MacArthur RH (1972) Geographical ecology. Harper and Row, New York

Macedo DS, Anderson AB (1993) Early ecological changes associated with logging in an Amazonian floodplain. Biotropica 25:151-163

Malloch DM, Pirozynski KA, Raven PH (1980) Ecological and evolutionary significance of mycorrhizal symbioses in vascular plants (a review). PNAS 77:2113-2118

Marquis RJ, Braker HE (1993) Plant-herbivore interactions: diversity, specificity, and impact. In: McDade LA, Bawa KS, Hespenheide HA, Hartshorn GS (eds) La Selva. Ecology and natural history of a Neotropical rain forest. Univ Chicago Press, Chicago, pp 261-279

Masera O, Ordóñez MJ, Dirzo R (1992) Carbon emissions from deforestation in Mexico. US EPA and Lawrence Berkeley Lab, Univ California, Berkeley

McKey D (1979) The distribution of secondary compounds within plants. In: Rosenthal GA, Janzen DH (eds) Herbivores: their interaction with secondary plant metabolites. Academic Press, New York, pp 56-133

McKey DB, Waterman PG, Mbi CN, Gartlan JS, Strusaker TT (1978) Phenolic content of vegetation in two African rain-forests: ecological implications. Science 202:61-64

McNaughton SJ (1985) Ecology of a grazing system: the Serengeti. Ecol Monogr 55:259-294

McNaughton SJ, Oesterheld M, Frank DA, Williams KJ (1989) Ecosystem-level patterns of primary productivity and herbivory in terrestrial habitats. Nature 341:142-144

Miller RM (1989) The ecology of vesicular-arbuscular mycorrhizae in grass- and shrublands. In: Safir GR (ed) Ecophysiology of VA mycorrizal plants, CRC Press, Boca Raton, pp 135-138

Montagnini F, Sancho F (1994) Net nitrogen mineralization in soils under six indigenous tree species, an abandoned pasture and a secondary forest in the Atlantic lowlands of Costa Rica. Plant Soil 162:117-124

Morton SR, James CD (1988) The diversity and abundance of lizards in arid Australia: a new hypothesis. Am Nat 132:237-256

Mulkey SS, Smith AP, Wright SJ. (1991a) Comparative life history and physiology of two understory Neotropical shrubs. Oecologia 88:263-273

Mulkey SS, Wright SJ, Smith AP (1991b) Drought acclimation of an understory shrub in a seasonally dry tropical forest. Am J Bot 78:579-587

Myers N (1980) Conversion of tropical moist forests. Natl Acad Sci Press, Washington DC

Nadkarni NM (1984) Epiphyte biomass and nutrient capital of a neotropical elfin forest. Biotropica 16:249-256

Nye PH (1955) Some soil forming processes in the humid tropics. IV. The action of soil fauna. J Soil Sci 6:73-83

Odum HT (ed) (1970) A tropical rain forest: A study of irradiation and ecology at El Verde, Puerto Rico. US Atomic Energy Comm, Oak Ridge TN

Oliveira-Filho AT (1992) Floodplain 'murundus' of central Brazil: evidence for the termite-origin hypothesis. J Trop Ecol 8:1-19

Olson DM (1994) The distribution of leaf litter invertebrates along a Neotropical elevation gradient. J Trop Ecol 10:129-150

Olson JS (1963) Energy storage and the balance of producers and decomposers in ecological systems. Ecology 44:322-331

Olson JS, Watts JA, Allison LJ (1983) Carbon in live vegetation of major world ecosystems. TROO4, US Dept Energy, Washington DC

Opler PA, Frankie GW, Baker HG (1976) Rainfall as a factor in the release, timing, and synchronization of anthesis by tropical trees and shrubs. J Biogeogr 3:231-236

Orians GH (1975) Diversity, stability and maturity in natural ecosystems. In: Van Dobben WH, Lowe-McConnell RH (eds) Unifying concepts in ecology. Junk, The Hague, pp 139-150

Paine RT (1966) Food web complexity and species diversity. Am Nat100:65-75

Parker GS (1994) Soil fertility, nutrient acquisition, and nutrient cycling. In: McDade LA, Bawa KA, Hespenheide HA, Hartshorn GS (eds) La Selva. Ecology and natural history of a Neotropical rain forest. Univ Chicago Press, Chicago, pp 54-63

Pfeiffer WC, De Lacerda LD (1988) Mercury inputs into the Amazon Region, Brazil. Environ Technol Lett 9:325-330

Prestwich GD, Bentley BL (1981) Nitrogen fixation by intact colonies of the termite *Nasutitermes corniger*. Oecologia 49:249-251

Prestwich GD, Bentley BL, Carpenter EJ (1980) Nitrogen sources for Neotropical nasute termites: fixation and selective foraging. Oecologia 46:397-401

Radulovich R, Sollins P (1991) Nitrogen and phosphorus leaching in zero-tension drainage from a humid tropical soil. Biotropica 23:231-232

Ramakrishnan PS (ed) (1991) Ecology of biological invasions in the tropics. Int Sci Pubs New Delhi

Reid WV (1992) How many species will there be? In: Whitmore TC, Sayer JA (eds) Tropical deforestation and species extinction. Chapman & Hall, London, pp 55-74

Rodin LE, Basilevich NI (1967) Production and mineral cycling in terrestrial vegetation. Oliver and Boyd, Edinburg

Sánchez PA (1976) Properties and management of soils in the tropics. Wiley, New York

Sanford RL (1989) Fine root biomass under a tropical forest light gap in Costa Rica. J Trop Ecol 5:251-56

Sanford RL, Paaby P, Luvall JC, Phillips E (1994) Climate, geomorphology, and aquatic systems. In: McDade LA, Bawa KS, Hespenheide HA, Hartshorn GS (eds) La Selva. Ecology and natural history of a Neotropical rain forest. Univ Chicago Press, Chicago, pp 19-33

Silver WL, Vogt KA (1993) Fine root dynamics following single and multiple disturbances in a subtropical wet forest ecosystem. J Ecol 81:729-738

Simms EL (1992) Costs of plant resistance to herbivory. In: Fritz RS, Simms EL (eds) Plant resistance to herbivores and pathogens. Univ Chicago Press, Chicago

Skole D, Tucker C (1993) Tropical deforestation and habitat fragmentation in the Amazon: satellite data from 1978-1988. Science 260:1905-1910

Snow DW (1962) The natural history of the oilbird, *Steatornis caripensis*, in Trinidad, W. I. II. Population, breeding ecology and food. Zoologica 47:199-221

Stark N (1971) Nutrient cycling II. Nutrient distribution in Amazonian vegetation. Trop Ecol 12:177-201

Stark N, Jordan CF (1978) Nutrient retention by the root mat of an Amazonian rain forest. Ecology 59:434-437

Stiles FG (1988) Altitudinal movements of birds on the Caribbean slope of Costa Rica: implications for conservation. In: Almeda F, Pringle CM (eds) Tropical rain forest: diversity and conservation. Cal Acad Sci Press, San Francisco, pp 243-258

Szott LT, Fernandes ECM, Sanchez PA (1991). Soil-plant interactions in agroforestry systems. For Ecol Manage 45:127-152

Tanner EVJ, Kapos V (1982) Leaf structure of Jamaican upper montane rain-forest trees. Biol J Linn Soc 18:263-278

Terborgh J (1977) Bird species diversity along an Andean elevational gradient. Ecology 56:562-576

Terborgh J (1986a) Keystone plant resources in the tropical forest. In: Soulé M E (ed) Conservation biology. Sinauer, Sunderland, MA, pp 330-344

Terborgh J (1986b) Community aspects of frugivory in tropical forests. In: Estrada A, Fleming TH (eds) Frugivores and seed dispersal. Junk, Dordrecht, pp 371-384

Terborgh J, Winter B (1980) Some causes of extinction. In: Soulé ME, Wilcox BA (eds) Conservation biology: an evolutionary-ecological perspective. Sinauer, Sunderland, MA, pp 119-134

Tilman D, Downing JA (1994) Biodiversity and stability in grasslands. Nature 367:363-365

Vitousek PM (1984) Litterfall, nutrient cycling and nutrient limitation in tropical forests. Ecology 65:285-298

Vitousek PM, Hooper DU (1993) Biological diversity and terrestrial ecosystem biogeochemistry. In: Schulze E-D, Mooney HA (eds) Biodiversity and ecosystem function. Springer, Berlin Heldelberg New York, pp 3-14

Vitousek PM, Sanford RL Jr (1987) Nutrient cycling in moist tropical forests. Annu Rev Ecol Syst 17:137-167

Vitousek PM, Walker LR (1989) Biological invasion by *Myrica faya* in Hawaii: plant demography, nitrogen fixation and ecosystem effects. Ecol Monogr 59:247-265

Walker LR, Brokaw NVL, Lodge JD, Waide RB (eds) (1991) Ecosystem, plant, and animal responses to hurricanes in the Caribbean tropics. Biotropica Spec Issue 23, 521 pp

Walter H (1973) Vegetation of the earth in relation to climate and the ecophysiological conditions. Springer, Berlin Heidelberg New York

Weaver PL (1989) Forest changes after hurricanes in Puerto Rico's Luquillo Mountains. Interciencia 14:181-192

Went FW, Stark N (1968) The biological and mechanical role of soil fungi. Proc Natl Acad Sci USA 60:497-504

Whitmore TC, Sayer JA (eds) (1992) Tropical deforestation and species extinction. Chapman & Hall, London, 153 pp

Wilson EO (1992) The diversity of life. Belknap Press, Cambridge, MA, 424 pp

Worbes M, Junk WJ (1989) Dating tropical trees by means of carbon-14 from bomb tests. Ecology 70:503-507

Wright SJ (1991) Seasonal drought and the phenology of understory shrubs in a tropical moist forest. Ecology 72:1643-1647

Wright SJ (1992) Seasonal drought and leaf fall in a tropical moist forest. Ecology 71:1165-1175

Wright SJ, van Schalk CP (1994) Light and phenology of tropical forest trees. Am Nat 143:92-199

Yoshinari T (1993) Nitrogen oxide flux in tropical soils. Trends Ecol Syst 8:55-156

Species Index

Topical Index

Ecological Studies
Volumes published since 1990

Ecological Studies
Volumes published since 1990

Springer-Verlag
and the Environment

We at Springer-Verlag firmly believe that an international science publisher has a special obligation to the environment, and our corporate policies consistently reflect this conviction.

We also expect our business partners – paper mills, printers, packaging manufacturers, etc. – to commit themselves to using environmentally friendly materials and production processes.

The paper in this book is made from low- or no-chlorine pulp and is acid free, in conformance with international standards for paper permanency.

Printing: Mercedesdruck, Berlin
Binding: Buchbinderei Lüderitz & Bauer, Berlin